ANT

ENCOUNTERS

PRIMERS IN COMPLEX SYSTEMS

ANT
ENCOUNTERS

Interaction Networks

and Colony Behavior

Deborah M. Gordon

PRINCETON UNIVERSITY PRESS
Princeton & Oxford

LIBRARY OF CONGRESS CATALOGING-IN-PUBLICATION DATA
Gordon, Deborah (Deborah M.)
Ant encounters : interaction networks and colony behavior / Deborah M. Gordon.
    p.   cm.   -- (Primers in complex systems)
Includes bibliographical references and index.
ISBN 978-0-691-13879-4 (pbk. : alk. paper)
1. Ants--Behavior. 2. Insect societies. I. Title. II. Series:
Primers in complex systems.
QL568.F7G635 2010
595.79'61782--dc22   2009031028

British Library Cataloging-in-Publication Data is available

This book has been composed in Adobe Garamond

Printed on acid-free paper. ∞
press.princeton.edu
Printed in the United States of America

1 3 5 7 9 10 8 6 4 2

I saw them hurrying from either side
and each shade kissed another, without pausing,
Each by the briefest society satisfied.

(Ants in their dark ranks, meet exactly so,
rubbing each other's noses, to ask perhaps
What luck they've had, or which way they should go.)

—Dante, *Purgatorio*, Canto XXVI

CONTENTS

Contents

LIST OF ILLUSTRATIONS

PREFACE

The ideas in this book grew in conversations and collaborations with many people, and I would especially like to thank (listed in the sequence in which I met them): John Gregg, Richard Lewontin, Simon Levin, Fred Adler, the late Lincoln Moses, Mike Greene, Susan Holmes, and Rodolfo Dirzo. I have learned a great deal from the questions and insights of the many people who have worked with me in the field. While writing this, I was inspired by the way that Helen DeWitt wrote about science writing, and the uncompromising character that she created, in *The Last Samurai* (Miramax, 2000). The manuscript was greatly improved by comments from or discussion with Erol Ackay, Gibson Anderson, Avis Begoun, Dan Edelstein, Paul Ehrlich, LeAnn Howard, Mark Longo, Lauren Ancel Meyers, Noa Pinter-Wollman, and anonymous readers. Adrienne Mayor's insight and guidance contributed to every stage of this book. I am deeply grateful to William Flesch for his wonderfully generous attention to the manuscript. All my thanks to Dennak for his encouragement and his perspective on interaction networks, and to my children, Sam and Eleanor, for coming along to the places I go to watch ants.

ANT

ENCOUNTERS

1

THE ANT COLONY
AS A COMPLEX SYSTEM

Ants are more than a hundred million years older than humans, and they cover the land surface of the planet. Probably people have always watched ants, and probably they have always asked the same question: How can ants get anything done when no one is in charge? Whoever wrote Proverbs 6:6 put it this way: "Look to the ant, thou sluggard—consider her ways and be wise. *Without chief, overseer or ruler*, she gathers the harvest in the summer to eat in the winter." The history of our understanding of ant behavior is the history of our changing views of how organizations work.[1]

There have been times when it was impossible to imagine an ant colony without a leader. The scientific study of ants began when natural history joined the rest of the emerging sciences in the eighteenth century. It was already clear that ants live in colonies, consisting of one or more reproductive females, while the rest are sterile females. Among bees, the reproductive female in a colony was called the 'queen,' and the females who do not reproduce called 'workers,' by Charles Butler in *The Feminine Monarchie, or the Historie of Bees*, in

1609.[2] These observations of bees were extended to ants in the eighteenth century by the French naturalist Réaumur. Like his contemporaries, such as Maeterlinck, writing about bees, Réaumur described ants as a group of subordinate laborers happy to serve their monarch. Although these names imply a hierarchy that in other times, both before and since Réaumur, was known not to exist, the names 'queen' and 'worker' have stuck. Two hundred and fifty years later, scientists have identified more than 11,000 species of ants, and they all live in colonies of females, with some sterile and some reproductive.

The 150 years that followed Réaumur's vision of ants as contented subjects of a benevolent queen brought worldwide political upheaval, raising questions of whether monarchy is the most natural form of society. This period (1750–1900), in which evolutionary biology was born, generated thinking about democracy, revolution, freedom, and cooperation, all of which influenced the ways we see the natural world, including ants. In a lively discussion in the Ecole Normale in Paris in 1795, year 3 of the French Revolution, Daubenton, a professor of natural history, argued that there is no royalty in nature—for example, the queen bee does nothing more than lay eggs. His colleague Latreille wrote in 1798 that the ants in the colony are not really subjugated workers; instead, the colony has "a single will, a single law" based on the love each ant feels for the others.[3]

Throughout the nineteenth century, colonial expansion put Europeans in contact with the stunning diversity of the tropics. Evolutionary biology and ecology began out of the effort, which is still under way, to explain this diversity. The idea of natural selection as the outcome of ecological processes, what

Darwin called "the struggle for survival," gradually became the basis for the scientific study of all organisms.

Skipping over many crucial discoveries about the life cycles, physiology, and natural history of insects generally and ants in particular, we could locate the beginning of contemporary scientific work on ants with the efforts of W. M. Wheeler.[4] Wheeler borrowed from Herbert Spencer the term "superorganism," comparing the ant colony not to a kingdom but to a single organism, with the queen and workers all acting as cells that contribute to the life of one reproducing body. Because ants do not make more ants, but instead colonies reproduce to make more colonies, a colony is in fact an individual organism in the ecological sense. As the gametes of different trees join, when pollen meets ovary, to make the seeds that produce new trees, so the reproductives of different colonies mate to produce new colonies.

With the colony as superorganism, the queen is no longer in charge, and we return to the puzzle of how such a system could be organized. This issue resonated with general questions about cooperation in animals raised in the early twentieth century by authors like Kropotkin, a Russian aristocrat turned anarchist. Do the ants work for the good of the colony, in the same way that cells work for the good of the body, and is this because evolution favors those who cooperate?

Despite all the transformations in our thinking about society, it is still very difficult for us to describe ant society without depicting it as hierarchically organized. Someone is always in charge. Either the bad guys are in charge, and the lowly workers feel oppressed and rebellious, or the good guys are in charge, and the lowly workers are happy. During the

Cold War, ants were models of a totalitarian society. In *The Book of Merlyn* by T. H. White,[5] Merlin transforms the young Arthur into an ant and sends him to work in a desolate tunnel with loudspeakers blaring allegiance to an ant Big Brother and walls plastered with signs reading "Everything not forbidden is compulsory." More recently, movies such as *Antz*, *It's a Bug's Life*, and *The Ant Bully* show the colony as a corporation with more or less disgruntled workers. These changes continue to be mirrored in scientific ideas about ants.

In the 1950s and 1960s, evolutionary biology took up an economic, free-market perspective with a vengeance. Anyone who did not see natural selection as promoting the gain, or profit, of the individual, was considered to be soft-headed and out of touch. Wheeler's ideas about the superorganism were scorned as soft-headed, "an appealing mirage," and "a panchreston of little relevance" (from a 1968 paper by E. O. Wilson).[6]

Wilson's pioneering work on ants, to which he brought his immense gift for making accessible to everyone the fascination of nature, was the starting point for modern research on social insects. Drawing on ideas from the nascent sciences of cybernetics, which led to the development of computers, he created a vision of colony behavior as a mechanical process, a "factory constructed inside a fortress."[7] Each component, the ant, was genetically programmed to do its task. An ant of a certain type would perform a certain task over and over, directed by its genes and responding to fixed chemical signals. Most ants see poorly, and they rely on chemical cues. An ant has many glands in its body that secrete chemicals. Working with Bert Holldobler, Wilson set out to find the meaning of each chemical—for example, one chemical may signal alarm, and another may mark a trail to be used by foragers. In their view, the chemical signals were the triggers for the ant's preprogrammed instructions to kick in. This view of the ant colony culminated in a set of mathematical optimization models. Oster and Wilson's 1978 book, *Caste and Ecology in the Social Insects*,[8] outlines how such a system would be tuned by natural selection to produce, in each species, just the right mix of ants to do each task as required by the environment. The queen was not in charge, but natural selection had stepped in instead, setting up the system in advance so that each ant does what needs to be done.

The idea of a perfectly adapted distribution of worker sizes was one answer to the question of how ant colonies could work without central control. More generally, this question is an instance of one of the fundamental puzzles of biology. In the early twentieth century, developmental biologists argued bitterly about another version of the same question: What determines the fate of cells in an embryo? All cells are formed from the division of one or two parent cells, so they all have the same genes. What then tells one cell to become liver and another to become bone? Does one organism, or one cell on its own, have inside it whatever determines its development, like the ancient idea of the preformed homunculus inside each sperm, or instead do the cells require interactions with each other to determine what they will become? It happened that the choice of systems used by developmental biologists in the late nineteenth century, sea urchins or frog embryos, helped to polarize this debate: the two sides had chosen organisms that differed greatly in how soon cell interactions become important, so that in one, isolated cells could go on to develop, and in the other, they could not.

Different outcomes of particular experiments, depending on choice of methods, have also shaped our ideas about ants. This was crucial to the course of my own work. I began to study ants as a graduate student in the early 1980s. The prevailing research program on ants at that time was set by the idea of the adaptive distribution of worker sizes in a colony, each type genetically programmed to respond to particular cues and perform particular tasks.

I was looking for an example of a system in which to investigate organization without hierarchical control. I was interested in embryonic development, but I chose ants instead because I learn best by watching, and it is a lot easier to watch ants than to observe cells as they divide and differentiate in a growing embryo.

Of the many species of ants, I chose harvester ants because one of my professors in graduate school at Duke, Fred Nijhout, told me of a well-known study by Wilson that concluded that for harvester ants, oleic acid is a necrophoric pheromone, causing any ant treated with oleic acid to be taken "live and kicking" to the refuse pile, or midden.[9] (There are many species of seed-eating or harvester ants, but in this book I use 'harvester ant' to mean my favorite, *Pogonomyrmex barbatus*.) I tried to repeat the experiment and found that ants did take bits of paper treated with oleic acid to the midden, but only at times when they were taking other things, like dead ants, to the midden. Harvester ants eat seeds, and many seeds contain oleic acid. When ants were taking other food into the nest, they also took oleic acid into the nest. Apparently, oleic acid functions either as garbage or as food, depending on what the workers that encounter it are doing.[10] When I sent my manuscript to Wilson to ask for comments, he told me that in his experiments, the ants treated with oleic acid had been chilled to keep them motionless. They were live but not really kicking, which might be why other ants took them to the midden.

The results of this study, the first chapter of my dissertation, took me in a direction orthogonal to the prevailing view of ant behavior. An ant's response to a chemical cue was not fixed, but depended on what the ant was doing. Then what determined what the ant was doing?

When I began to work on harvester ants, the closest place to find them was in a nature reserve next to the Army base at Fort Bragg, North Carolina. One day, after weeks of preparation, I had an experiment set up with little bits of paper soaked in oleic acid placed carefully around some nests. A group of soldiers landed their helicopter nearby to see what I was doing, producing a local gale that scattered the paper and the ants everywhere. I decided to find another place to work. I chose the Southwestern Research Station in southeastern Arizona, because all the other harvester ant species in the United States are in the southwest. When I went there for the first time, I fell in love with the desert. I grew up near the ocean and found the big sky and the desert wind coming across huge distances somehow familiar. I have returned there almost every summer since that first trip in 1981, to follow the same population of harvester ant colonies.

I've probably watched more ant colonies for longer than any other scientist, and for longer than most ant colonies have watched each other. Looking at the same harvester ant colonies week after week and year after year, I noticed that behavior changed. An ant's moment-to-moment response to a chemical cue depended on what it was doing right then.

A colony's response to its neighbors depended on what happened last week. Eventually, I realized that a colony's behavior changed over the years, as the colony grew older and larger. More and more, my questions were not about what ants do, but why ants and colonies change what they do. As I finished graduate school and moved into postdoctoral research, my work on harvester ants showed that individuals switch tasks in response to changes in the environment and interactions with other ants; an ant's behavior is not simply a set of fixed responses to chemical signals.

In his 1980 book, *Gödel Escher Bach*,[11] Douglas Hofstadter asked us to think of distributed processing systems as being like ant colonies. It turns out that this vivid analogy does not do justice to the ants. My experiments began to show that ant colonies display even more dramatic emergent behavior than the computer simulations that in the 1980s transformed engineering. Today the use of 'ant' algorithms is a thriving industry in computer science, artificial intelligence, and robotics, and the use of network theory informs our understanding of ant behavior. It is clear that ant colonies make collective decisions, similar to the ones that keep schools of fish and swarms of insects together. Such decisions dictate not only how ants move around, but also how colonies find the resources and maintain an environment in which to begin, grow, and reproduce. What exactly is the similarity between an ant colony and a computer program, or an artificial brain?

Recently, I gave in to several months of intense lobbying by my (then) 12-year-old son, and we drove to a dusty and remote former Air Force base in southeastern California for the DARPA Urban Challenge. DARPA (Defense Advanced Research Projects Agency) is the Pentagon's military research unit, and this event is part of an effort to encourage research leading to the development of vehicles that can navigate using moment-to-moment responses to their own sensors, without any need for remote control. There were 11 robotic vehicles in the contest. Each had to navigate a designated route through the streets of the Air Force base, including turns, parking, and changes of lane. Each of the robotic vehicles was followed by another vehicle, driven by a person, and other vehicles were driven around as well. The problem for each robotic car was to avoid bumping into any other cars, adjusting its movement in response to information from various sensors mounted on its exterior. The winner was the robotic vehicle with the fewest infractions of the California driving code.

To understand what a 'complex biological system' is, it helps to compare such systems with the collection of robots in the DARPA Urban Challenge. An engineer's view would emphasize the similarities. In this cybernetic view, each component, whether an ant, a cell in an embryo, or a neuron, has a mission. It accomplishes its mission using the input it gets from various sensors. To understand such a system, the problem is merely to figure out what sensory cues each component uses. For example, in the 1980s view of ants, the forager is told by its genes to go out on a food-collecting mission, pick up the scent of a particular pheromone trail, follow it, collect the food, and return home.

An ant colony and a group of robotic vehicles have in common some of the processes that determine not what happens when they interact, but whether they interact at all. For both ants and vehicles, whether any two meet depends on how they all move around. Even for the 11 vehicles in the Urban Challenge, it would be hopelessly complicated to predict where any two vehicles were likely to interact, and this was the reason to hold the event, a multimillion-dollar experiment, in the first place. Many small contingencies determined which of the robotic vehicles had to pass another, or when one had to wait for another to back out of its parking space before moving forward to park itself. In fact, there was one collision during the 6-hour event. One robotic vehicle edged into the right lane of a roundabout and then stopped. A second robotic vehicle came along in the left lane, but moved toward the right lane of the roundabout just as the first robotic vehicle moved forward, and the two collided. At this point, all the other robotic vehicles were offered the opportunity to pause, and they all did. So when the first robotic vehicle stopped in the roundabout, it set off a series of events that eventually affected all the other vehicles and changed their subsequent encounters.

There are always dense webs of contingency in systems of interacting parts. In the circulation of wind around the earth, the movement of molecules in a glass of water, the fluctuations of the stock market, the speed and reach of the Internet, or the six degrees of separation between any two people on earth, the effects of actions of one component ripple out to others. When one ant does something that involves another and changes the positions of those two ants, or how long they stay in the same place, this will eventually influence the positions of all the other ants. This is what makes complex systems complex. But the complexity of complex biological systems is not what makes living systems unique.

One way that living systems are unique, so obvious that it's easy to forget, is that they cause their own development and activity. For example, a basic difference between a collection of robots and an ant colony is that people make the robots, while ant colonies are made by other ant colonies. We intervene in biological processes, sometimes spectacularly, as when we clone sheep or administer vaccines, and in countless other ways so frequent and essential that we don't even think of them as interventions, as when we eat or plant seeds. But our interventions merely alter ongoing processes, such as the development of a sheep's egg or the distribution of nutrients around a body—processes that we do not make or control. Human designers are behind everything a robot does. In an ant colony, there's nobody behind anything.

The ways that ants respond to interaction allow them to do on their own what the robots can do only at our instruction. Interactions with other ants determine what an ant does, and what the ant does modifies its environment, including its interactions, and this in turn modifies its subsequent interactions—and the whole process runs itself. This is true of all living systems. In a developing embryo, each cell's fate depends on its interactions with other cells. Inside a cell, which genes turn on at a certain time is a response to changing chemical gradients and contact with other cells, and what genes are turned on determines what the cell produces and how it influences the local chemical environment, which feeds back to turn genes on and off.

In an ant colony, a forager leaves the nest on a mission, to collect food. Interaction sets up the forager's mission in the first place, since it is stimulated to leave the nest by returning foragers, and interactions it has later on, such as an encounter with an alarmed ant, can change its mission and send it back into the nest without food. For the robotic vehicles in the DARPA challenge, in contrast, interaction could not change the mission, which was to drive around without bumping into anything; it was merely a possible source of failure.

This book is an introduction to the ant colony as a complex biological system, but not a general introduction to ant behavior. It presents a single idea about ants: that the behavior of ant colonies arises from dynamical networks of interaction. The book starts with the moment-to-moment behavior of ants within colonies and then scales up, ending with the evolution of ants over more than a hundred million years. Chapters 2 and 3 are about colony organization and the role of interaction networks in regulating the behavior of colonies. Chapter 4 is on the function of colony size, which varies among species and also changes as a colony develops and grows. Chapter 5 discusses ant ecology, the relations of ant colonies with neighbors of the same species and with other species. Chapter 6 summarizes the little we know about the evolution of colony behavior, and chapter 7 concludes by outlining the prospects for general models of colony organization.

This book is based on the idea that an ant colony's behavior is guided by a pulsing, shifting web of interactions, in which the pattern of interactions is more important than the content. This idea came out of my early work, and inevitably here I draw most on my own work and the ants I know best. Ideas about collective behavior in general, and networks of interaction in particular, have begun to sprout everywhere, and there are many compelling examples of these ideas applied to ants that I left out to keep the book short. There are countless excellent studies of many other fascinating aspects of ant behavior that are not mentioned here at all.

The series that includes this book is produced by the Santa Fe Institute, which has done much to nurture complex-systems thinking. The Santa Fe Institute grew out of the realization that biologists, physicists, and chemists, all studying different systems, are discovering analogous processes. The big questions about ant colony behavior are the same ones we have to ask about the behavior, ecology, and evolution of any biological system, and the limits to what we know about ants are set as much by how we frame the problem as by the number of person-hours spent getting the answers. This book maps out an approach to learning more about ants.

2

COLONY ORGANIZATION

The Diversity of Ant Behavior

We don't know much about ant behavior for three reasons: there are so many different kinds of ants, it's hard to figure out what an ant is doing, and not many people have looked. Ants are an enormously diverse group. About 11,000 species of ants have been identified. Some people estimate that there are another 10,000 species, mostly in the tropics, that have not yet been found. Everyone knows that there are red ants and black ants; there are also yellow, green, and even blue ants, as well as orange, gray, and brown ones. They range in size from the little fire ant *Wasmannia aurapunctata*, barely 2 millimeters long, to the intimidating bola ants in the genus *Paraponera*, some more than 3 centimeters long, common in tropical forests.

Gordon Moore, the former CEO of Intel, once suggested that there were more ants on earth than human-made silicon chips. No one knows how many there are of either ants or chips, but there is no doubt that the number of ants is very, very large. There are ants on every continent on earth except

Antarctica, and ants live in every conceivable habitat and use an astonishing variety of nesting places and food. Although ants make up only about 2% of all insect species on earth, if you put all the living insects on a scale, about a third of the mass would be ants. One survey suggested that in the Amazonian rain forest, the weight of ants is about twice that of all the other land animals combined, including mammals, birds, reptiles, and amphibians.

Ants are about 140 million years old, much more ancient than the dinosaurs. The ancestors of ants were wasps, and the earliest ants already lived in colonies. The family of ants, Formicidae, has 15 subfamilies. The largest subfamilies are the Myrmicinae and the Formicidae. When an ant researcher meets an unfamiliar ant, she looks first at the petiole, the structure that joins an ant's thorax and abdomen, because this is the easiest way to distinguish the most common subfamilies. The myrmicines have petioles with two segments and the formicines have one.[1]

Of the approximately 11,000 known species of ants, only about 50 have been studied in detail. Ants are extremely diverse in ecology, where they live, what they eat, how they move around, how permanent their nests are, and how quickly colonies grow and reproduce. They must be equally diverse in behavior, because an animal's behavior produces its ecology. An ant that collects leaves to feed to fungus has to perform very different tasks from an ant that captures live prey.

Because ants are so diverse, it is misleading to generalize about ants. But all ants live in colonies, consisting of one or more reproductive females, the queens, and sterile, female workers. Even this basic plan has modifications, such as queens that inhabit the nests of other ant species, workers that

turn into queens, and other bizarre twists. The ants that you
see walking around are almost certainly sterile female workers, and back home in the nest, there is a queen, or queens.
The queens lay eggs to produce workers, daughter queens,
and males (although since males are haploid, workers of some
species can produce them from unfertilized eggs). Eggs grow
into larvae and then pupae, and the ant emerges from the
pupal case as an adult that will not grow any larger. Most of
the food that the colony brings in, especially the protein, goes
to feed the larvae.

Not many people have taken the time to watch ants carefully. In the nineteenth century, the English took their obses-
sions with birds and wildflowers around the world, to the
great benefit of ornithology and botany, but have you ever
heard of a local ant-watchers club?

If you watch an ant for a while, chances are that it will
appear to be aimless. Interpreting an ant's behavior is so difficult that often it's easiest to conclude that the ant is doing
whatever the observer thinks it ought to be doing. If the ant
is on the kitchen counter, it must be looking for food. If the
ant is near an ant of another species, it must be displaying its
brawn. You have to watch more than one ant to learn anything about their behavior, because of every five ants that
embark on a task, three will never manage to do it before you
get distracted or lose sight of them. The other two might turn
out to be doing something you never thought of.

Ant colonies perform many different tasks. A few tasks are
common to almost all species of ants. Ants leave the nest,
find food, and bring it back. They build the nest and repair
it. They feed and groom the larvae, and they move the pupae
around. Ants do an astonishing variety of other things as well.

To list just a few, ants weave nests out of silk spun by larvae; act as farmers by collecting leaves and feeding them to a fungus that the ants eat; protect aphids and eat the sugary liquid that the aphids secrete; kill plants by injecting them with formic acid; move from one nest to another, carrying all the stored food along; and conduct long raids across the forest, capturing all the insect prey they encounter along the way. Ants accomplish marvelous feats of engineering. Diane Davidson and others found that the Asian *Camponotus mirabilus,* which nests in bamboo, constructs a series of wicks and entrances to get rid of excess water.[2] The fungus-growing ant *Atta texana* and the fire ant *Solenopsis invicta* make elaborate underground tunnels to take foragers to their destinations. Joan and Gary Fellers found that several species of *Aphaenogaster* use pieces of leaves to soak up the juices of crushed dead insects and carry the food back to the nest, or to cover a newly discovered food source, thus hiding it from other scavengers while they go back to the nest to recruit nestmates to help retrieve it.[3]

We know of many amazing things that ants do, and still, anyone who watches ants is in for surprises. Trails of leaf-cutter ants, collecting leaves to feed their fungus gardens, look like a fleet of tiny sailboats, the ants carrying pointed bits of cut leaves high above their heads. But the first time I saw a leaf-cutter ant in the wild, it was dragging a caterpillar back to the nest. They aren't supposed to eat caterpillars.

Brian Fisher told me that once he was walking in the rain forest in the Peruvian Amazon, felt that he was being observed, and realized that it was an *Eciton* ant, perched on a branch and watching him go by. Most ants can't see, and visual tracking isn't generally part of the ant repertoire.

Robert Dudley and Steve Yanoviak discovered that *Cephalotes* workers can glide, moving deliberately through the air, not the usual means of travel for ants.[4] The discovery came about because Yanoviak got bored sitting in a platform high up in the canopy of the rain forest of the Peruvian Amazon, studying the biology of mosquitos. *Cephalotes* are large, slow ants with visors that resemble Darth Vader's. To entertain himself, Yanoviak threw ants over the edge of the platform and watched where they landed. He noticed that they seemed to show up very quickly back on the tree. This led to experiments with marked ants, which confirmed that the ants use their legs to steer as they fall, turning a vertical drop into an almost horizontal glide back to the tree.

Simon Robson reports that the Australian ant *Polyrachis sokholova* swims: the ant walks up to a puddle and then plunges right in, swimming neatly across in a kind of dog-paddle.

We've all heard about the marvelous efficiency of ants. Last summer, I spent a few hours in a forest in Mexico following a trail of another *Cephalotes* species, *C. goniodontus*, as they made their way from one tree to another. The ants went slowly up vines, down twigs leaning casually against other plants, and around tiny branches. The trail of ants never touched the ground and went a total of 38 meters to get from one tree to another 3 meters away. There may be a good reason, perhaps to avoid predators on the ground or other ants on the trunks of the trees, but I haven't figured out what it is.

Sometimes an ant does something completely out of character for its species. Phil Ward described to me a surprising burst of courage from an ant of *Pseudomyrmex apache*. These ants live on plants, mostly manzanita, in northern California. They usually vanish at the first sign of another ant species, retreating or darting around to the other side of the branch.

Ward was astonished to see a *Pseudomyrmex* worker confront another ant, the winter ant *Prenolepis imparis*. Instead of running away, the *Pseudomyrmex* worker burst out of its hole, clamped its mandibles around the *Prenolepis* ant, who was probably even more surprised than Ward was, and killed it on the spot. We will never know whether her nestmates gave her a purple heart or recommended therapy.

The best generalization we can make about ant behavior is that we don't know much about it, and ants continue to surprise us.

From Individual to Collective Behavior

All complex biological systems have in common that without central control, local interactions among the parts produce coordinated behavior of the whole. The main question is how the behavior of the component parts—the neurons in the brain, the cells in the immune system, the ants in the colony—produces the behavior of the whole system.

To learn how an ant colony works, the starting point is to identify a pattern in colony behavior and then to ask what the ants are doing to produce the pattern. To make this more specific, suppose that we look at an ant colony and see that right now some ants are foraging and some are cleaning out the nest. To explain this pattern, we then have to zoom in on particular ants. How does that ant come to be foraging and that ant working on the nest?

So let's begin with the question: What determines the task that an ant performs? Early work on social insects tried to answer this question using characteristics of individuals: this size or 'caste' does job 1, this does job 2, and so on. In this reductionist view, the behavior we see is due simply to the attributes of the components. The alternative to this perspective has many names, of which the best known are 'emergence' and 'self-organization'; other more recent ones are 'collective decision-making' and 'swarm intelligence.' Sometimes this alternative view is encapsulated by the expression that the behavior of the whole is greater than the sum of its parts. For ants, the alternative begins with the premise that what an ant does—for example, whether an ant goes out to forage—cannot be predicted from what we know about that ant alone.

The concept of emergence comes from work in the philosophy of science that analyzes explanations. The original definition of 'emergence' is that a phenomenon is emergent relative to a particular explanation if the explanation is not sufficient. For example, we cannot explain all the properties of water by spelling out the properties of its component parts, hydrogen and oxygen. What we know about hydrogen and oxygen does not tell us how water will flow. However, emergent phenomena do not occur by magic. We study emergent processes to find explanations for these phenomena, and if they are explained, then strictly speaking they will no longer be emergent.

In the late 1980s, models from statistical mechanics began to be used in artificial intelligence, beginning with the use of the Hopfield net to model neural networks. A 'Hopfield net' is a way to predict how changes in the orientation of electrons lead to phase transitions, such as from a liquid to a solid. Such models rely on a distributed process, in which any component responds the same way to particular conditions. 'Collective intelligence' is sometimes used to describe the phenomena described by this class of models. The 'swarm' models developed by Chris Langton at the Santa Fe Institute, and others, extended ideas about distributed processes to traffic, the stock market, and social insect colonies. Eventually, the metaphor of ant behavior came to be used to represent processes of interest to the artificial intelligence community, mostly solutions to the traveling salesperson problem about the quickest way to get through a series of steps, but also to investigate queuing delays and how interactions among units attempting to perform some task influence the rate at which they move through the task.

In 1977, Ilya Prigogine won a Nobel Prize in chemistry for his work on irreversible thermodynamics; he used the money to set up an institute in Brussels to find biological examples of irreversible thermodynamics. At the institute, one group, headed by Jean-Louis Deneubourg, noticed the analogies between the physical processes Prigogine studied and the process that initiates recruitment using chemical trails in ants.[5] To form a directed process, whether a stream of water or a trail of ants, all it takes is first some random movement and second a relation among the molecules, droplets, or ants such that one follows another.

This kind of process has come to be known as 'self-organization,' and following the lead of the Deneubourg group, who have developed a set of models to describe this, others have found many examples of this in ants.[6] Most are related to the formation of a trail in a certain direction or to the choice of a nest—if one ant follows another one, or follows a chemical it puts down, then a trail will form.

Guy Theraulaz and his colleagues have described a similar process that accounts for the way ants build tunnels or pile their dead.[7] If the ants are more likely to put a corpse where there is already a pile of corpses, and if ants are more likely to take a corpse to the nearest pile, then scattered corpses will end up in piles, and the more scattered the corpses are to begin with, the more different piles there will be.

Computer simulations and mathematical models show how networks operate and how we can use networks to make robots and computers do things for us. Models of self-organization and collective intelligence can use ant colonies as the inspiration for solutions to engineering problems that require groups of component parts to interact to perform a task—for example, getting a group of robots to move around on Mars taking photos, exploring links on a website to recommend products based on past purchases, or organizing a telecommunications system.

While engineers use mathematical models to figure out how to direct the operation of machines and computers, biologists use modeling for a different purpose, to learn how the natural world works. A model that specifies exactly how a natural system might work makes it easier to pinpoint the gaps in our understanding; discrepancies between the model predictions and nature help to specify what we don't know. Models are descriptions, sometimes very beautiful ones, of ways that we imagine the world to be. However, when something in nature—for example, the response of ant colonies to an experiment—acts like the entities in a model, this does not necessarily mean that the natural system is organized as the model is. There are many ways to describe any system; many different models could describe the same behavior.

I will not try to review here the many models that draw on ideas about ants, even those with data that show that ants behave like a model does. Many provide useful ways to investigate interesting processes. Almost all the models focus on a narrow slice of colony behavior, describing the relation among components or ants all performing the same task, usually recruitment to food. To demonstrate that these models explain the behavior of real ants, it would have to be shown that the ant behavior does not match many other models equally well, or that the model predicts some aspect of behavior not built into the model.

To turn from models back to real ants, the notion of 'emergence' reminds us that to understand an ant colony's behavior, we need to know more than the characteristics of each ant. In the reductionist paradigm, each ant is independently programmed to perform a task. If ant colonies really worked this way, it would be sufficient to know the program for each ant, and a complete list would fully specify the behavior of the colony. But we now know that the behavior of an ant also depends on its interactions with other ants. This doesn't mean that there is some colony soul that directs the ant's behavior. The behavior of the colony *is* the sum of the behavior of the ants, but the behavior of each ant depends on more than its own attributes. For some ant colony behavior, we are able to provide an explanation for apparently emergent behavior; we can specify how an ant's interactions with other ants and the rest of the world produce its behavior and how, when this process is aggregated for many ants, it produces the behavior of the colony.

To figure out how a colony works is to learn how the responses and interactions of individual ants add up to colony behavior. Colony behavior is dynamic, always changing because the colony's world is always changing. Stuff happens. An animal steps on the nest, or rain seeps in, and nest repairs are needed. There is a windfall of food, or there is a shortage. The changing environment continually shifts the numbers of ants required to perform each task, to repair the nest or collect food. The state of the colony changes too. Most of the food that a colony takes in goes to feed the larvae, so the amount of food needed must be tuned to the number of larvae. Since the queen lays eggs in seasonal or other pulses, there are pulses in the number of larvae begging for food, and so there are fluctuations in the amount of work that must be done to go out and collect food.

In the late 1980s, I introduced the phrase 'task allocation' to describe how the work of an ant colony is organized. Task allocation is the process that adjusts the numbers of ants performing each task according to the current situation, both in the world around the colony and inside the colony. The allocation of ants to various tasks is achieved by the whole colony, although no ant directs it or even understands what needs to get done. I began to use 'task allocation' in the hope that it would replace 'division of labor,' a perspective on colony organization discussed in detail in the next section. Division of labor, which implies specialized individuals, is only one of the many ways that a colony could accomplish task allocation, and it evokes a static procedure in which each individual is permanently assigned its place on the assembly line. I chose 'task allocation' to focus on the collective outcome of colony organization, however it is accomplished, and to emphasize that colonies shift their behavior in response to a changing world.

Division of Labor

Keeping in mind that no ant directs the behavior of others, what determines which task an ant performs and when it performs it? The film *Antz* beautifully embodies many widespread misconceptions about ants. In the film, each ant's task is chosen by bureaucrats. Cherubic little larvae are carried to a booth. A harried clerk looks at each one for a second, calls out "worker" or "soldier," and then stamps the larva, thus determining its task for life. The bureaucrats in the film take the place of genes to depict the notion that each individual is genetically programmed to do a certain task. This perspective on social insect behavior is usually characterized by the phrase 'division of labor.'

The idea of 'division of labor' is familiar: workers on an assembly line each perform a different task. Adam Smith introduced the idea of division of labor in 1776 in *The Wealth of Nations*[8] to explain why it would be better for a society if each individual specialized in certain jobs. In Smith's ideal village, one man makes candles, while another is a farmer. Smith argues that this is better than if each person tries to do all tasks. Smith's explanation for the benefits of division of labor includes some observations about people that could not possibly extend to ants, such as the idea that when a man is forced to do a task over and over, he will invent a machine to do it better. Adam Smith suggests two basic advantages of division of labor in people: first, we get better at a task when we practice it, and second, we differ in the ways our talents and inclinations equip us for certain tasks. There is little evidence that ants learn to get better at their tasks by doing them. The proponents of the 'division of labor' perspective on ants did not argue for this first advantage. Instead, they emphasized the second of the advantages that Adam Smith proposed, that certain individuals are better suited than others for particular jobs. In introducing this perspective, E. O. Wilson considered mainly the minority of ant species in which there are workers of more than one size. The idea was that each worker would perform the task for which its size is best suited—for example, the small workers would carry small objects, while the large workers would attack intruders.

'Division of labor' is a misleading way to refer to the process that determines an ant's task. There is little evidence that ants are specialized to perform certain tasks. Discussion of this has been confused by equating what *ought* to happen, if colonies were behaving optimally according to a particular scheme, with what *actually* happens. The original models of division of labor in ants described an ideal: colonies could produce the optimal distribution of sets of workers, each set of the size best suited to perform a particular task. This was an attractive idea, but even the most elegant argument about what ought to happen does not demonstrate that it does happen.

Early empirical work on division of labor sought to show that ants of different sizes are specialized to perform different tasks. Much of this work was done in several *Pheidole* species and in *Atta cephalotes*, all species with workers that differ strongly in size. The size of an ant is fixed once the ant emerges as an adult; a small worker never grows to be a larger one. Each size worker was called a 'caste,' evoking human social systems in India in which a person's position in society is considered to be determined at birth. The goal of these studies was to show first that workers of a particular size perform a particular task and second that the task is particularly suited to the ant's size.

In these early studies, the observer noted the task of ants of each size. The results showed some overlap among tasks performed by a particular size of ant. Statistical techniques were developed to decide whether the tasks performed by one size ant were really distinct from those performed by another. Some studies concluded that they were.

However, even in species with different sizes of workers, an ant's behavior changes if conditions change. Removal of the ants of one size causes the others to switch task. For example, Wilson found that in many *Pheidole* species, the removal of minors, the smaller ants who tend to perform brood care, caused majors, the larger ants, to switch to brood care.[9]

Although some studies suggested that in species with workers of different sizes, workers of a certain size tend to perform certain tasks, the next step, showing that each size performs the task it is best suited for, proved to be much more difficult. Wilson attempted this with laboratory colonies of the leaf-cutter ant, *Atta cephalotes,* which cuts leaves to feed to the fungus garden inside the nest; the ants eat the fungus.[10] He isolated groups of ants by size and then measured the rate at which they cut leaves into pieces and the amount of oxygen they consumed. The premise was that the ants that could cut the most leaves per amount of oxygen consumed were the most efficient. There were two kinds of leaves, soft rose petals and harder rhododendron leaves. He found that the sizes of ants that usually cut rose petals were not necessarily the most efficient, but the sizes of ants that usually cut rhododendron leaves were the most efficient, cutting more leaves into bits per volume of oxygen consumed.

The conclusion in this study was that *A. cephalotes* has evolved to maximize the efficiency of cutting hard vegetation but not soft vegetation. But this conclusion begs many ecological questions. Maybe a group of ants of a single size in a laboratory colony cuts leaves differently from ants working in an intact colony, or one in the field. Maybe it is not important to the colony's survival and reproduction how fast or how well the colony cuts leaves, or maybe a few slower ants are just as effective as one super leaf-cutter. Maybe size matters in some way that has nothing to do with leaf cutting. For example, tiny ants sometimes hitchhike, riding along on larger ones. Don Feener showed that the small ants chase off a parasitic phorid fly that lays its eggs in ants' heads.[11] More recently, Charles Yackulic and others found that hitchhiking by tiny ants has other functions as well. The hitchhikers collect sap leaking from the edges of the leaves that the larger workers are carrying, and remove fungal parasites from the leaves before they are carried into the nest.[12] So the evolution of various sizes of workers of fungus-growing ants might depend less on their efficiency in cutting leaves, and more on the ability of a certain size of ant to deal with a certain size of parasite, or the benefits to the fungus, the ants' food supply, of a thorough cleaning of the leaf fragments. Our judgment about what seems most efficient may not match up with the real action of evolution.

The argument about division of labor was that natural selection should produce the optimal distribution of ants within a colony. If the ants of a certain size are the best foragers, and the colony needs to devote 40% of its effort to foraging, then, it was argued, 40% of the ants should be of the ideal foraging size. Thus, if natural selection produces the optimal distribution of workers to do each task, then the distribution of workers should reflect the needs of the colony. In different conditions, colonies would have different needs.

For example, Sam Beshers and James Traniello reasoned that in the fungus-growing ant *Trachymyrmex septentrionalis*, the size distribution of colonies in Florida, where the warmer climate makes it possible to collect food throughout the year, should differ from the size distributions of colonies in the colder climate of New York, where more large workers might be needed to collect more vegetation during the shorter growing season.[13] They did find that colonies in the Long Island population had more large workers than in the Florida one, although in both places the distribution of worker sizes changed as colonies grew older and larger. However, Beshers and Traniello found no task specialization according to worker size—large workers don't collect more food. They suggest that the distribution of worker sizes might arise from selection for rapid colony growth and that worker size is not related to division of labor.

Else Fjerdingstad and Ross Crozier examined how much ants diverge in size in 35 species of ants.[14] They found that workers were likely to differ more in size in species in which queens are much larger than workers, suggesting that the developmental process that leads queens to be larger also affects the growth of workers. Again, this would mean that the size distribution of workers is not related to division of labor.

Many attempts to show that ants of a particular size perform best at a certain task never led to any strong conclusions. The idea that a particular size is best suited to a particular task was the basis of the argument that ants *should* have evolved to specialize on certain tasks. In the vast majority of ant species, all ants are the same size; variation in worker size occurs in only 44 of 263 genera of ants. While ants of different sizes could be especially well suited to perform certain tasks, it is difficult to see why task specialization would help when ants are all the same size. Unlike people, ants do not seem to get better at a task by performing it over and over.

Whether or not we think that there ought to be division of labor, there is abundant evidence that ants change tasks. This means that we have to shift the focus of evolutionary questions about colony organization. Instead of asking how the colony evolves to have a static, optimal distribution of specialized workers, we need to ask how the colony evolves the moment-to-moment regulation that gets the necessary numbers of workers into each task according to current conditions.[15]

As everyone knows, when there is a picnic, there will be ants. Task allocation determines how the colony gets more ants to the picnic, and which ants will go. To explain colony behavior, we ask how the actions of individuals, none of which can assess the overall situation, allow the colony to adjust the numbers of ants performing each task.

Ants Switch Tasks

I have studied which ant performs which task in the red harvester ant *Pogonomyrmex barbatus*.[16] These seed-eating ants are common in the deserts and dry grasslands of the southwestern United States, Mexico, and South America. Four tasks are performed outside the nest: foraging, patrolling, nest maintenance work, and midden work. *Foragers* travel away from the nest in streams reaching 10 to 30 meters from the nest and then fan out and search for seeds, which they bring back to the nest to be processed and stored. The *patrollers* are the first ants to leave the nest in the morning. They search the nest mound and foraging area, and choose the day's foraging directions. It is the return of the patrollers that stimulates the foragers to begin their work for the day. *Nest maintenance* workers carry out the dry soil that collects inside the nest during the excavation and repair of underground chambers. *Midden workers* manipulate and sort the refuse pile, or midden.

These four exterior tasks are probably performed by the oldest 25% of the colony. There is no way to determine an ant's age except to mark it when it first emerges from the pupa as an adult, and since the youngest ants are deep in the nest, they are impossible to mark without destroying the nest. So we can only infer the relation of task and age from the sequence of tasks an ant undergoes, by assuming that the ants are older when they reach later steps in the sequence.

An ant may change its task from one day to the next, if conditions change. To learn about how harvester ants switch tasks, we marked ants working outside the nest with dots of colored paint. This is relatively easy to do with harvester ants, because they are about a centimeter long, with large heads. Once the paint dries, it seems to be odorless, because the other ants do not respond any differently to marked and unmarked ants. Probably the cuticular hydrocarbons that carry the colony's signature odor are spread afterward over the dry paint by grooming, and since the ants can't see much, they do not perceive any difference between a bright green ant and an unmarked brown one.

When we marked ants according to task, we found that from one day to the next, if nothing drastic happens, an ant will continue to do the task it was doing the previous day. (This is not true for younger, smaller colonies; more on this in chapter 4.)

I found that ants switch tasks if more ants are needed to perform a particular task.[17] Not all transitions are possible. If more foragers are needed, workers of the other three tasks will switch tasks to forage. If more patrollers are needed, nest maintenance workers will switch tasks to patrolling. If more nest maintenance workers are needed, they must be recruited from the younger workers inside the nest. Then, once a worker becomes a forager, it does not switch back to any other task. Thus, foraging acts as a sink, while the younger workers inside the nest, who will be recruited to nest maintenance if needed, act as a source.

Midden workers bring in pebbles from the area around the nest to cover the mound. They reinforce the colony odor in these pebbles.[18] I don't know where ants enter midden work in the sequence from nest maintenance to patrollers to foragers. The numbers of midden workers are much smaller than those of any other task group. We are investigating exactly how ants use the midden material as a repository for the colony's chemical signature, and how much and how often the odor is distributed on the mound surface. When we know more about this, we will be able to create situations that would require more ants to be recruited to do midden work, so we will be able to find out which task groups new midden workers come from.

Once an ant's task is determined, how does the ant know how to do it? We don't know the answer, for ants or, really, for any other animal. We refer to 'instinct' or, more recently, 'hard-wiring' or 'programming,' but this metaphorical language cushions almost complete ignorance. There is a pervasive fantasy that genes are little packages of instructions that tell us and other animals how to behave. However, we know that in fact what genes do is determine the production of proteins. The expression of genes, which determines which proteins are currently being produced, is transient and context-dependent. Most important, even when we can track which proteins are manufactured when and by which genes, we still do not know how to explain behavior as a function of these gene products.

How an ant manages to perform any particular task is a fascinating puzzle. To understand how the ant knows what to do, we also have to understand what is the tolerance for error, or the usual range in how well the ants perform each task. It's important to remember that whatever the ant is doing, it's not rocket science. As a nest maintenance worker, a harvester ant picks things up inside the nest and puts them down outside; later, as a forager, it picks different things up outside and puts them down inside. In both cases, it has to pick up the right things and put them down in the right place, and it has to get back and forth. Each ant's performance doesn't have to be perfect, and it usually isn't. Instead, enough ants have to perform well enough, often enough, for the colony to get enough food and not have the entrance blocked by garbage.

Age Polyethism

Ants change task as they get older.[19] It is generally assumed that this happens in ants as it does in honeybees. When a honeybee changes from working inside the nest to going out to forage depends very consistently on the bee's age. However, honeybees have been artificially selected for at least 10,000 years to forage in predictable ways so that we can use them as pollinators. This intense artificial selection has probably reduced the variation in how long it takes for a bee to become a forager.

In contrast with honeybees, how an ant's task changes with age is clearly extremely variable. Not only does the progression differ among species, it differs among individuals within a colony and among colonies of the same species, depending on conditions. A general pattern, however, is that younger workers stay inside the nest, working on brood care and nest construction, and then move to work outside when they are older.

Ants often start out working near the place where they eclose from the pupal case and emerge as adults. This means that the first task workers perform will be brood care, because they emerge from the pupal case among the other pupae. Later, they may leave the brood chamber and find themselves in a place where another task is being done, such as sorting seeds or repairing the nest. Nigel Franks gave the name 'foraging-for-work' to the processes that shuffle an ant from one location to the next, so that eventually it finds itself near the nest entrance and stimulated to work outside.[20] This idea combines the notion of shifting task as the ant ages, called 'age polyethism,' with the notion that an ant's task is determined more by its location than by any particular characteristics of the ant itself.

Although the idea that an ant moves from inside to outside tasks over its lifetime is part of the dogma of ant biology, there are surprisingly few data from any ant species showing that the transition from inside to outside work occurs consistently, with ants of the same age performing the same task. There are many articles that suggest this, but few that demonstrate it. For example, of 21 articles cited by Holldobler and Wilson's review of this topic in *The Ants*[21] in support of the idea that ants move from inside to outside work as they grow older, only 2 had any data showing this transition.[22]

The only way to know an ant's age is to mark it at birth and follow it over time. Because marking newly eclosed ants deep in a nest in the field would be so severe a disturbance that it would probably influence the sequence of tasks, studies of the relation of task and age are done in the laboratory.

In a laboratory study, Marc Seid and James Traniello found that in *Pheidole dentata*, an ant does a greater variety of tasks as it gets older.[23] Which tasks an older ant performed also differed greatly from one laboratory colony to another. Seid and Traniello suggest that as an ant grows older, it reacts to a larger set of stimuli, and thus ends up participating in a greater variety of tasks. Mario Muscedere and others in James Traniello's lab showed that older workers were more effective than younger ones at caring for the larvae and queen.[24] They suggest that we see younger workers engaged in brood care more than older workers only because it takes many attempts for a younger worker to accomplish the same amount of feeding and grooming that an older one can do in fewer bouts.

In laboratory experiments with the carpenter ant *Camponotus floridus*, Frederic Tripet and Peter Nonacs observed whether younger or older ants did brood care, an inside task, or foraging, an outside task.[25] They found that younger ants were more likely to do brood care and older ones to forage, but when they removed one age group, the other one would switch. Thus, the colony's need for foragers or brood tenders, exacerbated in the experiment by removing the ants normally performing that task, had a stronger influence on an ant's task than its age. Philip McDonald and Howard Topoff had similar results with *Novomessor albistetosus*.[26] When all age groups were removed except the oldest, the older ants went back to tending the brood and the queen inside the nest. When the older ants were removed, contact with larvae stimulated foraging in the younger ants.

We followed marked harvester ants in laboratory colonies to see if they worked inside the nest when young and outside when old.[27] The results showed remarkably high variation in the ages that ants started working outside the nest, especially considering that all the colonies were kept in very similar, stable conditions in the laboratory. It seems likely that the transition from nest work when young to outside work when old occurs fairly consistently in the field, because we find that ants change from nest work inside the nest to foraging outside and don't switch back. However, since we have little data on the ages of workers in the field, we don't know for sure if foragers as a group tend to be older than nest workers. If it's true that ants move from inside to outside work more consistently in the field than in the unnaturally stable conditions of the laboratory, this suggests that the requirements imposed by variable external conditions in the field have a very strong influence on the sequence of tasks an ant performs.

The progression from one task to another is the product of interacting factors. The most obvious of these is the spatial dynamic to which 'foraging-for-work' refers. Workers tend to emerge inside the nest; in fact, this is true by definition because whatever the nesting arrangement, we call the place where brood are found and where pupae become adults 'inside the nest.' A second factor is the recent history of colony needs. For example, if past events created a need for workers to repair the inside of the nest, ants might remain inside longer and thus be older before they come out to forage. In contrast, if past events created a need for new foragers, they will have been recruited from younger ants, and overall, foragers will be younger. A third factor is the current demands of the colony and the environment. For example, when workers are removed, other workers change tasks; if many foragers are lost to a predator, inside workers might become foragers when they are younger. Fourth, some individuals are consistently more active, throughout their lives, than others, and these more active individuals are more likely to leave the nest. At any time, the average age of ants performing a given task will be the combined outcome of all of these forces. It would be surprising if the outcomes were not extremely variable.

What Ants Respond To

How colonies act depends on what ants can perceive and respond to. Ants react to two kinds of external information: changes in the outside world and interactions with each other.

The most important sensory mode of ants is olfaction. Ants use their antennae to perceive odors from objects they touch with their antennae or from the air. Much of the rest of this chapter is about olfaction, but ants use other senses as well.

Like many other insects, ants are sensitive to vibration. They hear vibration with the subgenal organ in their legs, which senses vibration in whatever they are standing on. Some ant species produce a sound and a vibration by stridulation, rubbing a rough patch on the end of the thorax against another on the top of the abdomen. Detecting vibration is crucial for many ant species. For example, *Azteca* ants live in hollow chambers inside branches of the tree *Cordia alliodora* throughout Central America. The ants defend the tree from insect herbivores, attacking anything that eats its leaves. The vibration produced by a caterpillar as it lands on a leaf, or moves slowly along, brings out hordes of *Azteca* to chase it off the plant.

In many species of ants, the ants can distinguish light from dark but do not see much more than that. Some ants can see distinct shapes. If you have spent a lot of time watching ants without inspiring any interest at all on their part, it is both shocking and a little gratifying to wave your hand over a nest of red wood ants, because the ants look up at you! These large, orange-red ants (*Formica rufa* group) are widespread in northern Europe. Their nests are mounds of pine needles, and the nests of old colonies can be enormous, up to 2 meters high. Their close relatives are found across northern and mountainous areas of North America. All the species in this group build nests out of piles of twigs or pine needles, and they too will look back at you when you wave.

There is still much to learn about the sensory capabilities of ants. There is some evidence that ants can perceive magnetic fields. Perhaps this helps ants navigate. So far, we know that ants of different species find their way using various combinations of chemical cues, visual landmarks, and the direction of sunlight.

Inhabiting a world in which smell and touch dominate, a huge range of events can claim an ant's attention. Ants respond not only to picnics, but also to countless stimuli that we can perceive and many that we can't: the smell of food, the smell of a person's breath, the smell that signals whether another ant is a nestmate, the presence of the aphids or scale insects the ants tend, light entering a dark place, changes of temperature, water, an obstacle or the absence of a passageway, their own brood, whether the floor of a chamber is flat, tiny vibrations made by the movement of other insects, and many more.

Ants are constantly reacting to each other. Interactions between ants are usually chemical or tactile: an ant responds to the odor of the ants it meets or to a chemical the other ant has emitted, or it responds to the impact or vibration caused by another ant.

Perhaps the best known chemical cue used by ants is trail pheromone. It is familiar because many of the ant species we see most often, the ones that take advantage of resources provided by humans—like the ants on your kitchen counter—use trails. A scout ant finds food and puts small amounts of a chemical on the ground, or on your counter, in a line or trail back to the nest. Other ants at the nest then follow the trail back out to the food, in some species laying more trail for others as they return.

Argentine ants, in contrast, put down trail pheromone all the time, both coming and going. The ants move around using a highway system of trails that connect the colony's many nests. By offering ants food on the other side of a soot-covered bridge from laboratory nests, and examining the tiny lines in the soot made by ants' abdomens as they streaked the ground with pheromone, Serge Aron and others showed that Argentine ants lay trail as they go, constantly reinforcing the trail.[28] The ants also make patrolling forays off the trail. Every summer, I get calls from people who are puzzled to find a heap of dead Argentine ants in their freezer. The ants are attracted to something, presumably an odor, in the rubber lining of freezer doors. No ant finds the freezer and goes back to recruit the others; once an ant goes in the freezer, it is doomed. But since the ants lay trail wherever they go, the ants that are attracted to the freezer all lay a trail on their way to it, and this is reinforced by more curious but equally doomed ants. Since Argentine ants are enormously abundant in many parts of the world, this procedure must lead to food more often than to the untimely death of many ants.

Not all ant species use trail pheromones. For example, in many species of seed-eating ants, the ants do not normally follow pheromone trails. In the red harvester ants that I study, patrollers early in the morning choose the direction foragers will take, but their chemical signal extends only a tiny way, about 20 centimeters, guiding the foragers toward a much longer trail that can extend for tens of meters.[29] Later, each forager returns to the place it last found a seed. The ants collect seeds that are scattered around the desert and that one ant can easily bring back by itself, so they do not need to recruit groups to work together to collect large food sources.

Many ant species use chemicals to signal alarm. Alarm pheromones are volatile, dispersing quickly in the air. Alarmed ants often run around in circles, sending out more pheromone that gets more ants running around in circles, so there is a spreading wave of alarmed ants. Alarmed ants are likely to react aggressively to whatever they meet as they dash around.

An ant's body has about 15 glands (depending on the species), each of which secretes a different substance. We do not know the function of most of these chemicals, nor how much they are used in combination. Our ignorance is partly due to the limits of our technology for measuring tiny amounts of chemicals, which make it difficult for us to detect what ants are putting out as the chemical is emitted. This inhibits our ability to interpret the ants' chemical signals in context. In contrast, we can easily hear the calls of many primates, and an observer who can hear the calls and see the resulting behavior has the relatively straightforward task of figuring out which response goes with which sound. For ants, however, most experiments involve putting down an extract of a gland and watching the response of the ants. Since we don't know how much an ant would put down, or exactly when, or where, our experiments tend to be crude, and this can make for misleading results. Imagine a Martian investigator, working to decipher the English language, performing a similarly crude experiment. The Martian, trying to discover the meaning of the word 'ant,' drops a large billboard bearing the word 'ANT' out of its spaceship into a busy city intersection at rush hour. The Martian then sees panicked people, hears sirens, and soon realizes the people are attempting to blow up the spaceship. It would be a mistake to interpret this reaction as the meaning of the word 'ant.' The main difficulty of research on chemical communication is working out ways to examine responses to particular chemicals in realistic quantities and appropriate contexts.

Task Allocation

How does a colony, with no one in charge, get the right numbers of ants to perform each task when needed? To answer this question is to outline a process that combines the actions of individuals into a collective outcome. In the same way, the answer to the question "How does a traffic jam form?" is a description of a process in which the actions of many individuals have a collective outcome.

An ant does not perform according to instructions—from some inner program, or from other ants of higher rank. Ants use local information, such as chemical communication, but they do not tell each other what to do.

To understand how task allocation operates, we have to learn how the behavior of individuals, in the aggregate, produces the behavior of colonies. It's obvious that what an ant does depends on what it perceives in its environment, because we see ants react to stimuli we can identify. If ants did not react to new food in the environment, there would be no ants at picnics. It takes a little more work to reach the conclusion that what an ant does also depends on its interactions with others. The first line of evidence was the experiments begun in the 1980s showing that when ants performing one task are removed, ants of another task change their behavior.[30] This suggests that the task that an ant performs somehow depends on the other ants present, but does not demonstrate that ants respond directly to other ants. It could be instead that the ants react to the work piling up that the ants that were removed would have done. For example, suppose that the ants husking seeds are removed. If other ants switch to seed husking after the removals, this could be because they perceive that the seed-huskers are missing, but it could also be because they perceive that the unhusked seeds are starting to get in the way.

To test whether an ant's behavior depends on its interactions with others, I did a series of experiments with harvester ants in which I changed the conditions affecting one group of workers.[31] I wanted to see if this would change the behavior of another group of workers. For example, I put out piles of toothpicks near harvester ant nest entrances early in the morning, when the nest maintenance workers clear out the nest. This led to a large increase in the numbers of nest maintenance workers, who moved the toothpicks to the edge of the mound, where everyone else just ignored them. My question was whether the increase in nest maintenance work would change the numbers of ants performing other tasks. It turned out that when the numbers of ants performing nest maintenance increased, the numbers of ants foraging decreased. This was true for several activities that I interfered with: the numbers performing one task changed when the numbers performing another task were altered by my intervention.

These results showed that as well as switching tasks, ants must be making moment-to-moment decisions about whether to perform their task actively. Not only do interactions determine whether an ant will be a forager today, but also whether, once it is a forager, it will go out to forage right now or instead wait around inside the nest. For example, the decrease in foraging when nest maintenance was increased could not be because the foragers had switched to nest maintenance. My other experiments with marked ants showed that once an ant becomes a forager, it does not switch back to nest maintenance, and new nest maintenance workers are recruited from the younger workers inside the nest. This means that the foragers stay inside the nest when nest maintenance is increased.

An ant's behavior depends both on what it perceives in the world around it and on its interactions with other ants.

The cues that influence its behavior must be local, because chemicals can't be detected at a distance, and the cues must be rather simple, because ants aren't very smart. How does an ant translate these local cues into action? How does this add up to colony behavior? The ant's reactions determine which task it does, and whether, at a particular moment, it is active.

3

INTERACTION NETWORKS

The rest of this book explores the idea that an ant responds to its pattern of interactions, and so the behavior of ant colonies is the result of networks of interactions. Because of the many applications of distributed processes in engineering, and the pervasive role of the Internet in our lives, everyone now knows what a network is. But in 1989, when I first began to suspect that an ant's behavior depends on its experience of interactions with others in a network, it was difficult to find familiar terms to describe such patterns. I started out using both 'network' and 'encounter pattern' interchangeably. I used 'pattern,' the most general term I could think of, because I didn't yet know which feature of the pattern was used by the ants. If interaction *rate* was important, this might explain my most puzzling results of the previous few years: older, larger colonies are more stable than young, small ones (see chapter 4). The reviewers of my first papers on ant networks and encounter patterns were bitterly opposed to the idea that ants could use the pattern of contact itself, rather than any particular message conveyed during contact, as a source of information. But over the last 15 years, it has become clear that many biological systems are regulated by networks of interaction among the components, from genes to individuals. Now, collective behavior is widely studied in social insects, even by some of the people who at first resisted the idea. (Perhaps this acceptance is due to a high rate of encounter with network ideas.)

'Network' has a technical meaning and an ordinary meaning. The technical meaning is a map of the links produced by a particular relation among components. The relation of sending and receiving e-mail creates the network of the Internet. The relation of meeting and knowing the names of people creates the network of acquaintances. The relation of having ancestors in common creates the network of an extended family. Each participant, each sender and receiver, or each family member, is a 'node.' Networks can vary in shape and in structure, depending on how the components are connected and how many connections there are per node. Because the idea of a network can apply to so many different kinds of system, and because a network can take so many different forms, it is a general way to describe how interactions among parts produce the behavior of a system.

In its ordinary meaning, 'network' evokes a fairly regular array of connections, like chicken-wire or a honeycomb. But to speak of a network of interactions in an ant colony (or a brain or an immune system) is not to say that the interactions are patterned in any simple or regular way. It is colonies, not ants, that behave in a predictable way. In a particular ant species, colonies perform a standard sequence of tasks each day. Colonies respond to disturbances in much the same way each time. A colony's behavior transforms in predictable ways as it grows older and larger. One colony's relations with its neighbors look much like another's. Although there is variation, and noise, there are clearly patterns in the behavior of an ant colony.

The patterns or regularities in ant colony behavior are produced by networks of interaction among ants. The networks of interactions are complicated, irregular, noisy, and dynamic. The network is not a hidden program or set of instructions. There is no program—that's what is mind-boggling, and perhaps it is why, at the beginning of the twenty-first century, there is so much we do not understand about biology. It is very difficult to imagine how an orchestra could play a symphony without a score. It takes an effort to avoid slipping into thinking that there is an invisible score hidden somewhere.

What Happens at Network Nodes

An interaction network is a set of relations among the participants. We all participate in many such networks: within families and at work, on the Internet, among users of cell phones. Networks differ in what happens at each node and by whether information or some substance is transferred.

Antennal contact is the crucial interaction in many ant networks. If you watch ants, any ants, you will notice that they sometimes walk up to each other and touch antennae, or one ant touches another's body with its antennae. During this interaction. one ant smells another and can tell whether the other is a nestmate. Contact with chemical cues left by other ants on the ground or on a plant is another important kind of interaction.

An ant uses its recent experience of interactions to decide what to do. The pattern of interaction itself, rather than any

signal transferred, acts as the message. What matters is not
what one ant tells another when they meet, but simply *that*
they meet. An ant operates according to a rule such as, "If I
meet another ant with odor A about three times in the next
30 seconds, I will go out to forage; if not, I will stay here." The
rules are actually more probabilistic than that—more like, "If
I meet another ant with odor A about three times in the next
30 seconds, the probability that I will go out to forage will
increase by about 10%; if not, it will go down by about 20%."
Brains use interaction networks in an analogous way. A neuron's function depends on its recent experience of interactions
with other neurons, and out of these patterns of interaction
emerge thought and memory.

The function of interaction networks is to transfer information, using 'information' in the technical sense to mean
an event that produces a change of state. Let's suppose that
we can define every ant's state with two variables, one that
describes the task group it currently belongs to, like foraging
or brood care, and one that describes whether it is currently
active or currently inactive. Interactions transfer the information that can change the ant's state, from inactive to active or
from one task to another.

The patterns that regulate the transfer of some substance
can also be considered networks. Many involve distinct
groups of ants. For example, consider the network that feeds
a colony. The ants performing one task, foraging, bring in
the food; another task group is the ants that process the food,
by husking the seeds or digesting the insect prey or feeding
the leaves to the fungus; a third task group is the ants that
get the food to the larvae. Such steps in the transfer of a substance, which Chris Anderson and Francis Ratnieks call "task

partitioning,"[1] produce the flow of food into the colony, the flow of waste out of the nest, and so on. Much more is known about the flow of food, water, waste, and chemical signals in other social insects, especially honeybees, but also wasps and termites, than in ants.

The Pattern of Interaction Is the Message

An ant moves around in a world of frequent contact with its nestmates, while it is out performing a task or when it goes back to the nest. That contact sustains the activity of the ant. The ant begins to do something, and its current experience of contact sets the probability that it will continue, or stop, or start something else.

An early example of the effect of interaction rate on task allocation is Wilson's 1985 result that when the smaller workers, or minors, of *Pheidole pubiventris* species are removed, the larger ones, or majors, switch to perform brood care.[2] This is the outcome of a simple rule of interaction: when majors met minors near the brood pile, they turned away. When minors were removed, there were fewer minors around. This meant that majors were less likely to meet minors and instead more likely to encounter other majors, and so they did not turn away, but instead stayed to help with the brood. Although Wilson explained this process as "between-caste aversion," it could also be interpreted as an example of task allocation that depends on interaction rates.

When ants interact by touching antennae, one ant perceives the cuticular hydrocarbons of another. Cuticular hydrocarbons are greasy fatty acids that are spread by grooming over the hard outer surface of the ant's body. Many insects have a layer of hydrocarbons on their cuticle, and these are important sources of identity and mating cues. Cuticular hydrocarbons are greasy, and probably their original function is to keep the insect from drying out. Now these chemicals also function as labels. In social insects, each colony has a characteristic cuticular hydrocarbon profile, a set of many different chemicals in quantities particular to a certain colony. This odor makes it possible for one worker to identify whether another one is of the same colony.

A graphic demonstration that ants use cuticular hydrocarbons comes from work on Argentine ants, who are unusually reluctant to fight with ants from another colony. Even though fighting between colonies is rare, Argentine ants can be induced to fight when they are fed foods that produce changes in cuticular hydrocarbon profiles. Dangsheng Liang and Jules Silverman discovered this when a lab technician combined two trays of ants, of which one had been fed the German cockroach *Suppella longipalpa* while the other had not.[3] The researchers were astonished to see many ants fighting. This led to a series of experiments that showed that the cuticular hydrocarbons of the cockroach were quite similar to those of Argentine ants except for one aberrant component. Ants that eat the cockroach, or even come into contact with it, acquire this new component from the cockroach into their own hydrocarbon profiles, and this addition provokes the ire of other Argentine ants.

There are huge regions within which the Argentine ants do not fight, in areas of California and the Mediterranean coastline, where these ants, originally from Argentina, have invaded and become established.[4] It seems that whether Argentine ants fight depends on the similarity of the food they eat and the impact of their food on their hydrocarbons. The boundaries of the regions in which fighting is rare may be determined mostly by the distribution of McDonald's and Taco Bells on the California coastline, and perhaps by more refined distinctions in the contents of garbage cans used by Argentine ants along the French Riviera.

The interactions that regulate foraging in harvester ants use cuticular hydrocarbons. Foraging begins in response to interactions between patrollers and foragers.[5] Colony activity begins early in the morning, when a small group of patrollers leave the nest mound. This is probably stimulated by the warmth of the first touch of sunlight in the nest entrance; nests in the shade tend to begin patrolling later. The first patrollers meander around the foraging area, and eventually return to the nest. Foragers are stimulated to leave the nest for the first time in the morning by the return of the patrollers. If patrollers are prevented from returning, the foragers do not emerge. What guarantee do the returning patrollers provide? If a patroller can leave and return safely, without getting blown away by heavy wind or eaten by a horned-lizard predator, then so can a forager. The patrollers also put down a chemical on the nest mound that shows the foragers which direction to take when they leave the nest;[6] this is discussed in chapter 5.

Whether a forager leaves the nest to begin the day's work depends on its interaction with returning patrollers. Foragers recognize patrollers during antennal contact, using the task-specific cuticular hydrocarbon profile.

We found that in harvester ants, not only do all nestmates share a colony-specific chemical profile, but in addition the

hydrocarbons differ within a colony according to task. When
an ant works outside, its odor changes. Annie Bonavita-
Courgourdan and others had shown earlier that carpenter
ants (*Camponotus vagus*) working inside and outside the nest
differ in cuticular hydrocarbon profile.[7] Diane Wagner and
others in my lab found that in harvester ants, hydrocarbon
profiles differ among task groups because the conditions in
which an ant works change its hydrocarbons.[8] Ants performing tasks that involve long periods outside the nest, such as
foraging, have hydrocarbon profiles with a higher proportion
of n-alkanes than ants performing tasks centered inside the
nest, such as nest maintenance. Exposure to warm, dry conditions increases the proportion of n-alkanes in an ant's hydrocarbon profile.

Michael Greene and I did an experiment in the field that
showed that forager activity is stimulated by contact with
patroller hydrocarbons.[9] Knowing that the foragers will not
go out unless the patrollers return, we kept the patrollers from
returning to the nest, by picking them up. Someone crouch-
ing by the trail (wearing gloves because the ants sting) can
easily swoop in, grab patrollers as they walk back to the nest
and put them in a plastic box. We then replaced the patrollers
with patroller mimics: little glass beads coated with extract
of hydrocarbons from that colony's patrollers. We dropped
glass beads into the nests of colonies whose patrollers had
not returned. Glass beads treated with patroller hydrocarbon extract stimulated foraging. Glass beads treated with
hydrocarbon extract from another task, nest maintenance, or
treated only with solvent as a control, did not stimulate foraging activity. Contact with beads that smell like a patroller is
enough to stimulate the foragers to leave the nest.

Although ants assess cuticular hydrocarbons when they interact, the cuticular hydrocarbons from one ant do not signal or instruct another ant to do a task. Instead, cuticular hydrocarbons merely provide a way for an ant to identify the task of the ants it meets, and thus to track its rate of encounter with ants of a particular task.

The rate at which patrollers return is crucial to stimulate foraging.[10] Glass beads that smell like patrollers do not stimulate foraging unless they are introduced at the correct rate. Foraging begins when patrollers return at a rate of about 6 per minute or 1 per 10 seconds, and glass beads must be introduced at a rate of 1 per 10 seconds or foraging does not begin. One of the few ways we have ever succeeded in getting ants to do our bidding was to drop in beads coated with patroller extract at the rate of 1 per 10 seconds before foraging began. We were able to trick colonies into starting to forage earlier.

Once foraging begins, the number of ants that are out foraging at any time is regulated by interactions with foragers coming back with seeds. Foragers back inside the nest after their last trip are stimulated to leave the nest again by the return of successful foragers bringing in food.[11] We learned this by taking away foragers as they returned to the nest. If we took ants returning with food, the rate at which other foragers go out slowed down (even accounting for the foragers that were missing because we had put them in a plastic box to prevent them from returning.) In contrast, if we took away the few ants that return without food, there was no effect on the rate at which foragers went out. Recently, Michael Greene and I learned that inactive foragers respond to returning successful foragers because they respond to the combination of two odors: the hydrocarbon profile of foragers and the food

itself. Either odor alone, that of a forager without food or of
food without a forager, is not sufficient to stimulate inactive
foragers to leave the nest.[12]

The rate at which foragers return with food indirectly mea-
sures food availability. Each forager travels away from the nest
with a stream of other foragers and then leaves the trail to
search for food. When it finds a seed, it brings it directly back
to the nest. The duration of a foraging trip depends largely on
how long the forager has to search before it finds food.[13] So the
rate at which foragers bring food back to the nest corresponds
to the availability of food that day. The rate at which successful
foragers are returning increases when food is abundant and easy
to find.

Surprisingly, a change in the rate of forager return translates
very quickly, within minutes, to a change in the rate at which
more foragers go out.[14] When we removed returning foragers
for 3 minutes, decreasing the rate at which foragers return,
the rate at which foragers leave slowed immediately, within
2 minutes. It's hard to see why the colony has to respond so
quickly, when it is foraging for seeds that lie around in the
soil for months. It may be that this fine-tuning of foraging
effort to food availability helps the colony to conserve energy
and water lost while foraging. But another possibility is that
the rate of response is determined simply by the ants' short
memory, on the scale of 10 seconds.

Harvester ant colonies adjust their behavior to food supply
using a network of brief antennal contacts. Each ant reacts
only to the rate at which it meets other ants of a certain task.
An ant assesses the task of another using its task-specific
hydrocarbon profile. Thus, the ant is using simple rules such
as, "I'm a forager—when I meet a few patrollers at a rate of

about 1 per 10 seconds, I'll go out for the first time," or more precisely, "I'll be much more likely to go out for the first time." If the rate of interaction with patrollers slows down— perhaps the ant meets mostly nest maintenance workers, or meets patrollers much less often than once per 10 seconds— the forager continues to wait inside the nest.

Once the forager goes out, it searches until it finds and brings back a seed. It comes into the chamber just inside the nest entrance and drops the seed on the floor, relying on other ants to do what my children wish I would with their dirty socks: pick up the seed and take it farther down into the nest for processing. The forager then waits inside the nest entrance until it meets enough foragers returning with food to be stimulated to go out. In this way, local interactions among ants add up to adjust the foraging activity of the colony to the availablity of food.

Another well-studied interaction network is used in nest choice in *Temnothorax*. These tiny ants nest in twigs and acorns, and colonies move frequently from one nest to another. When the colony needs to move, ants investigate new sites and are more likely to stay in viable sites. Then scouts recruit the rest of the colony to the site where they encounter enough ants. Stephen Pratt showed that scouts use their rate of antennal contact with the ants at the new site to determine whether a site has satisfied enough ants.[15] If the density of ants at the new site is high enough, the scouts will go back to the old nest and recruit the rest of the colony to move. The scouts merely assess the numbers, not the satisfaction of the ants in the new site, but because satisfaction is linked to the probability that an ant will stay in the new site, numbers of ants in the new site are correlated with its quality.

In some species, trail pheromones are used in an interaction network. Many trail pheromones are volatile. When an ant detects trail pheromone, the ant that deposited it was there recently. Wilson showed that in fire ants, the amount of trail pheromone is linked to the abundance of the resource.[16] The more ants find the food, the more ants lay trail, and the stronger the chemical trail. The quantity of pheromone on the trail is a measure of how many ants have recently been sufficiently excited about the food to put down a chemical. A strong trail acts as a high rate of interaction.

In principle, interactions, the events at the nodes of ant networks, can provide positive or negative feedback. In harvester ants, antennal contact with midden workers generates positive feedback. An ant that is not performing midden work is more likely to switch to do midden work when its rate of contact with midden workers is high.[17] Of course, this raises the question why all ants don't eventually end up doing midden work. There must be some other factor, such as the amount of hydrocarbon in the midden material, that leads to negative feedback. Another example of positive feedback is that foragers are stimulated to leave the nest when successful foragers come in. This response to interaction increases the numbers of active foragers, who will return to the nest to interact more, and thus intensify foraging. What dampens the response is first the limited pool of available foragers—numbers of foragers cannot increase indefinitely because the colony would run out of foragers. As you will see later in this chapter, ants may also respond to an increase in interaction rate by avoiding contact, and in this way, interaction provides negative feedback that slows the accelerating intensity of response.

Rate and Memory

Interaction rhythms produce colony behavior as a result of the relation of two rates: the rate at which interactions occur and the rate at which ants respond. It's obvious that the rate of interaction is important: as encounters stimulate ants to respond, the frequency of encounters determines how quickly ants change their behavior. Suppose that you had a robot that moved only at your command. It would move more slowly if you issue commands once a day than if you do so every second. In the same way, the colony can respond to interactions only as fast as ants interact. The rate at which individuals respond matters too. The ants' response rate depends on how long they can remember, or more precisely, on the duration of the interval over which an interaction can continue to influence an ant's behavior. Suppose that you tell your robot to start blinking rapidly. If you give a command every second, and the robot remembers your command for the whole second, you could get your robot to blink all the time. But if the robot remembers your command only for a millisecond, it will blink, then do no blinking until another second goes by, blink again when it gets the next command, not blink for a while, and so on.

We can measure how often interactions occur, but it is much harder to figure out how long an ant remembers an interaction or continues to respond to it. In fact, we know little about how an ant experiences its rate of interaction. Our results on harvester ants are consistent with a threshold response, one that requires some level of sensory input to be effective. For example, Rob Page and others have shown that honeybees respond to the odor of sugar once the odor exceeds a sensory threshold (which varies among individual bees).[18] An ant's perception of interaction rate might be based on a system that requires a certain rate to reach a threshold. It could work like this: Each time a forager meets a patroller, the encounter triggers a response in the forager that decays over time. For example, the response could be to stimulate the release of some neurotransmitter. The response seems to decay very quickly, in about 10 seconds. Each response to an interaction adds to the probability that the forager will leave the nest. When encounters occur rapidly enough, each successive encounter occurs before the effect of the previous one has fully decayed and increases the probability that the ant will forage. Eventually, if there are enough encounters per unit time, the responses add up to exceed some activation threshold, and the ant goes out to forage. Thus, if we introduce beads that smell like patrollers, once every 10 seconds, foragers leave the nest. But if one encounter occurs long after the last one has decayed, the probability of foraging goes back to the initial value. The ant has forgotten that the last encounter ever happened, and it has to start over again accumulating interactions until it reaches the activation threshold. Thus, if we introduce the same beads, once every 45 seconds, each encounter occurs too late, after the response to the last one has decayed, and the foragers don't leave the nest. A threshold response like this is analogous to the process that determines whether a neuron fires.

A rapid decay in a harvester ant's response to interactions would explain why colonies react so quickly to a change in the rate at which successful foragers return. If the probability that an inactive forager leaves the nest depends on the rate at which it meets other ants, but it responds only to very recent encounters, then overall the colony will respond only to very recent shifts in the rate of returning foragers.

The capacity to react to the interval between two events seems to be widespread in social insects. Two examples from studies of the flow of substances show that individuals assess intervals between interactions. Bob Jeanne's studies of nest-building in *Polistes* wasps show that the interval between loads of water brought to the nest influence nest construction behavior.[19] Tom Seeley showed that honeybee foragers respond to the interval they must wait to have their nectar unloaded; when the wait is long, bees take longer to go out again.[20] This makes sense for the colony (although the bee doesn't think about it) because the wait is longer when the nectar storers are busy and there is less need for nectar. These studies show that like ants, bees and wasps assess the duration of very short intervals on the scale of seconds.

We can consider an ant's capacity to respond to something that happened in the past to be an instance of the more general process of memory, remembering an association or an event from the past. We know very little about ant memory. In laboratory experiments, Fabienne Dupuy and others trained carpenter ants with sugar solution.[21] The ants could remember for 5 minutes which smell was associated with sugar. Zhanna Reznikova's laboratory experiments with red wood ants suggest that ants might remember for minutes or hours, and communicate to others, which way to turn in a maze to arrive at a food source.[22]

A harvester ant forager seems to remember locations. It returns to the same site over and over each day on successive foraging trips.[23] It is possible that the ant remembers the site all day, but it could also accomplish the behavior we observe with a much shorter memory, on the scale of minutes rather than hours. The average foraging trip is about 20 minutes. The forager leaves the nest, proceeds directly to the site where it first found food that morning, and then searches around for another seed. Once it finds food, it goes directly back to the nest and drops its seed just inside the nest entrance. If the forager then encounters many other foragers returning with seeds, within seconds it will leave the nest on its next trip. Since the ant often takes only a few minutes to return to the nest, leave again, and get back to the same site to search, it has to remember the location of the site only for these minutes. Then when it gets back to the site, it could refresh its memory and start over.

Maybe ants can remember a location for much longer than a few minutes. A harvester ant forager seems to remember from one day to the next which direction it took the previous day.[24] Michael Greene and I did some experiments to figure out how patrollers tell foragers which direction to take. We found that patrollers put a chemical from a gland in the abdomen, the Dufours gland, on the nest mound. This short chemical trail, only about 20 centimeters long, directs the foragers to leave the nest mound in a certain direction. The foragers may then travel a futher 20 meters in that general direction. When we prevented the patrollers from putting out a chemical cue in a certain direction, the foragers did not go in that direction. To find out whether a forager's memory of the direction it took the previous day is important, we marked foragers according to the direction they traveled one day, and the next day we prevented the patrollers from putting a

signal anywhere on the mound. In the absence of any patroller signals, the foragers went in the direction they went the
day before—so somehow, a forager can remember which way
it went the previous day. However, even though foragers are
capable of remembering yesterday's direction, it is clear that
usually patroller signal overrides forager memory, because colonies change trails from one day to the next. If foragers were
relying on memory with no contribution from the patrollers,
they would go only where they went the day before, and the
trails would not change direction from day to day.

Ant species seem to differ greatly in how long they remember their colony-specific odor. David Fletcher showed that in
fire ants, several days without a queen was needed before ants
were willing to accept another queen.[25] One way to explain
this is that it took several days for the ants to forget the queen's
smell. However, it is not possible to rule out the possibility
that the ants were responding to the presence of the queen's
odor, spread around from ant to ant by grooming, which
after one to three days without the queen decreases enough
that ants change behavior and accept a new queen. Christine
Errard's experiments with mixed-species groups of *Formica
selysi* and *Manica rubida* suggest that even after a year, ants
remembered the hydrocarbons they were exposed to in the
first three months of their lives.[26]

The memory of individual ants could be extended through
communication to produce colony memory over a much
longer interval. Rainer Rosengren's work on red wood ants
(*Formica rufa* group) suggests that ants pass on a cultural tradition.[27] Foraging trails lead from the nest into trees where
the ants drink the honeydew excreted by aphids feeding in
the trees. Foragers tend to use the same trails over and over.

In the winter, the ants huddle together deep inside the nest. Many ants die over the winter, and pupae kept over the winter emerge as adults in the spring. Rosengren found that in the spring, an older ant, which survived over the winter, leads a young ant out on its preferred trail. Then the old ant dies, and the young ant adopts that trail. The older forager must remember to go on the same trail at the end of the winter as it did in the autumn, and the young forager must remember, from one day to the next, to go on the trail that the older one showed it the day before—but the colony remembers the trail for decades.

A repeated process can have an outcome that looks like it is the result of memory. Auguste Forel, the nineteenth-century Swiss myrmecologist, observed *Polyergus*, slave-making ants that collect brood of another species, *Formica fusca*, and bring it back to their own nest, adding workers to the *Polyergus* slave-making colony. Forel noticed that if brood remains in the raided nest, the slave-makers may return many times to collect it, but if there is no brood left to steal, the slave-making colony does not return. He attributed this to the raiders' memory of how much brood was left behind.[28] Howard Topoff studied the same group of species in the southwestern United States. He found that a single scout returns to a previously raided nest, and if there is brood remaining, goes back to recruit its nestmates.[29] Forel must have missed this single scout. Repeated checking by a scout replaces long-term memory of the amount of remaining brood, and the outcome is the same: if there is brood left, the ants go to collect it, and if not, they leave the nest alone.

Another example of colony memory that arises from repetition is described in chapter 4. When neighboring harvester ant colonies use overlapping foraging areas one day, they usually forage in different directions on the next. This occurs not because any ant remembers not to go back to the place where it met the neighbors, but because repeated checking by patrollers leads colonies to avoid repeated encounters.

To understand how ants use interaction networks, we will need to learn more about how often ants interact and how long their reaction to each interaction can last.

Individual Variation

The simplest way to think about the structure of interaction networks is to consider all the ants to be the same. In a distributed network, differences among individuals don't matter; any individual performs the same function at a particular node. What is fascinating about ants is the tension between ants as unique individuals, each with purpose and agency, and as identical, like cells in a tissue. This contradiction appears over and over, in different forms, in stories about ants. In the *Iliad*, the Myrmidons, an army of selfless, fearless soldiers, were ants that had been turned into people by Zeus to repopulate an island decimated by the plague. The soldiers were antlike, despite their human form, in their dedication to the army and disregard for self. In T. H. White's Cold War story of King Arthur's visit to an ant colony, the ant colony is a totalitarian hell. The ants are unique individuals but are treated as if they are not. The same theme is the premise for the film *Antz*, which begins with an ant, voiced by Woody Allen, wondering to his therapist why he feels somehow different from the others and oppressed by the conformity of the colony.

We can see that the ants of any real colony are not all the

same. Pierre Jaisson and others have found that ants differ in activity level.[30] Some ants move around more than others, and these are characteristics that persist throughout the lifetime of particular individuals. From a network perspective, differences among individuals in how much they interact with others could lead to large differences in their impact on others. For example, among humans, one man infected with HIV had a huge effect on the spread of AIDS in San Francisco because he had sexual contact with so many people. The effects of individual differences in contact rate have not yet been studied in ants.

Individuals might differ not just in the rate of contact, but in the function or impact of contact. Small differences in cuticular hydrocarbon profile probably function in the behavior of many ant species. An ant's hydrocarbon profile, and thus the odor other ants perceive on it, is constantly changing. Harvester ants of different tasks differ in hydrocarbon profile, because the conditions of an ant's work affect its smell. The longer an ant is outside, the more n-alkanes in its hydrocarbon profile. This suggests that on the first day it forages, an ant's signature odor is not yet as fully forager-like as it will be after a few weeks. Does this matter? Perhaps the effect of an interaction with a forager depends on its seniority on the job. If so, the interaction network shifts over time with respect to the role of particular individuals. It might not matter in principle to Esther whether she meets Rosemary or Priscilla, but if Rosemary smells more like a forager than Priscilla, a meeting with Rosemary will have a different effect on Esther from a meeting with Priscilla.

If some ants are more active than others, they might do more work or influence the way that work is organized—for example, by encountering more ants. To find out about this in harvester ants, we marked hundreds of foragers with unique marks. Each ant got three dots of paint, one each on its head, thorax, and abdomen. With five or six colors, we are able mark large numbers of ants, and they are not only decorative but also easily distinguished. (The pioneer for our use of a multicolor system was an undergraduate, Jennifer Chu, who has both amazing hand–eye coordination and a photographic memory for which combinations had been marked; she is now training to be a surgeon.)

We wondered whether more intense foraging was due to the activity of particular, especially eager foragers.[31] When we looked at the number of foraging trips each marked ant makes, it was clear that there are a few ants, about 10% of the foragers, that make many trips, while the rest make only a few. To find out if the ants that make many trips were particular ants especially likely to achieve heroic feats of foraging, we did a two-day experiment. On the second day, we removed the 20 or so marked ants that had made the most trips the day before. The result was that removals made no difference. When we took away the ants that had done the most foraging, other ants moved in to fill the gap, and new ants then made many trips. So if the star foragers disappear, others quickly step up to take their place. This means that an ant who is not a star forager one day can easily become one the next. This suggests that there is nothing different about the particular individuals who made more foraging trips.

Whether a forager turns out to be one of the few that makes many trips may depend only on where it happens to find its first food of the day. Once it finds food somewhere on its first trip, that is where it returns. Most of the time an ant spends on a foraging trip is spent searching for food; they travel so fast that travel time accounts for very little of the trip, even if the ant goes far away. When an ant's first stop of the day is a place where food is abundant, it doesn't spend much time searching, and it ends up making many quick trips, because on each trip it returns to a place where food is abundant and search time is low. If an ant settles on a place where food is scarce, all its trips that day will be long because it spends a long time searching. So the apparent differences among foragers can arise not because some ants are especially earnest, but because locations differ in food supply. We found a slight day-to-day correlation in the number of trips a forager makes, but not a strong enough one to suggest that some ants are better than others, day after day, in finding places with enough food to make search times low. In harvester ants, at least, whether colonies forage more or less does not depend on the activity of certain forager heroines.

Another way that individual differences could matter is if ants of different kinds are required to join a team. Some tasks are performed by teams of workers. In the weaver ants (*Oecophylla*), nests are made of leaves with edges glued together by larval silk. Some ants hold the leaves together; some hold the larvae above the seam. The ants performing each task are not specialized on that task. All members of the team are needed, but Ethel might hold a larva one time while Samantha holds the leaves together, and another time Ethel might hold the leaves together while someone else holds the larva. There is no reason why it would be better if Ethel holds the larva, as long as somebody does it.

In species with different sizes of workers, teams may require ants of certain sizes to participate. Claire Detrain and others showed that when ants of *Pheidole pallidula* encounter a prey too large for one ant to carry, small ants will recruit larger ants to help carry it.[32] When prey is easy to carry, ants lay only a weak recruitment trail. When prey is hard to budge, ants lay more trail. The larger majors are more likely to respond to higher concentrations of trail pheromone. This pattern of interactions, between ants and the trail pheromone laid by other ants, generates teams that retrieve large prey. The larger the prey, the more likely the team is to include some large ants. In this case, the team might not need a particular large ant, say Ethel, but it needs some ants that, like Ethel, are large.

Species Differences

Comparing the interaction networks of different ant species would provide a quantitative way to identify differences among species in colony organization. It's clear that ant species differ greatly in the tempo of their reactions. Rapid response is what is most alarming about the red imported fire ant, the invasive species that has become established throughout the southeast United States. It seems that any contact with a nest is immediately followed by zillions of stinging ants all over you. This may seem extreme to people used to the more placid ants of the temperate zone, but the experience is all too familiar to anyone who has walked through a tropical forest. Stinging ants seem to rain down on you from all sides. Many tropical ant species are specialized mutualists with plants that grow structures to house ants, and in turn the ants defend the plant from herbivores. These plant-ants take defense seriously, and they respond very rapidly to the touch of a caterpillar on a leaf or to the touch of a human on a tangle of vines that leads to the host tree.

At the other extreme, an example of ants that are slow and set in their ways is the red wood ants of northern Europe. They create huge nests of pine needles, mounds that can be 2 meters high, and the nests persist for centuries, occupied in turn by many generations of colonies. Foraging trails, into trees where the ants tend aphids, last for many years, and ants are firmly committed to particular trails. Rainer Rosengren, Lotte Sundstrom, and I created pools of transient foragers by moving marked ants from one nest to another nest of the same colony. Once ants had chosen a direction at the new nest, even bait would not persuade them to change.[33]

Ant species differ in how quickly individuals move, but speed is not really what determines the differences among species in tempo. The rapidity of response generally depends on the rapidity of the interactions among ants. Differences among species in the speed and intensity of the colony's reactions come from differences in the rate at which the network is ticking, how often the ants interact, and how quickly and how much they respond. All of the variation among species in interaction networks begins with differences in the shape of the paths that ants use to move around.

The shape of an ant's path determines how much ground it covers. Vincent Fourcassie tracked the paths of single *Formica* foragers and showed how they adjust path shape to the food source.[34] The same principle extends to a group. When many ants search together and convey information to each other about food, then the pattern formed jointly by their paths determines how well the whole colony covers ground.

How ants move around also determines how often they meet. In the paths of fire ants shown in figure 3.1, there seem to be more turns and more interactions near food. If an ant interacts more when it is near food and turns more when it interacts, ants that arrive near the food from somewhere else will meet other ants more and turn more. This will keep ants that arrive near the food close to it and more likely to find it. In general, if an ant reacts to its rate of encounter by changing the way that it moves, then each encounter will change the probability of future encounters. A simple example is the observation by Blaine Cole that when acorn ants are kept in a small dish, there are cycles of movement.[35] The cycles occur because one ant meets another and causes that ant to move, so it jostles the next one, and so on until they get to the edge of the dish and the wave bounces back. Another example is the effect of alarm pheromone, which can create a wave of agitated ants. When ants are alarmed, they start running around in circles, spreading the alarm to other ants, who then run around in circles too.

Fred Adler and I modeled the relation of path shape and information transfer in groups of ants. We found that how quickly a group of ants can find something and how quickly information about it spreads through the colony are both enhanced by increased colony size and by straighter paths. These two effects interact. Smaller groups must use straighter paths to get the same result as a larger group.[36]

Later, I found that Argentine ants behave as our model predicted. When Argentine ants are at high density, their paths become more convoluted—they search round and round in

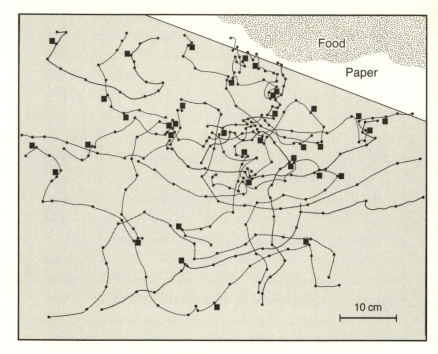

Figure 3.1. Fire ants interact more around food. The figure shows the paths of fire ants (*Solenopsis invicta*) as they move around a box in the laboratory. The solid squares show where two ants had a brief encounter, touching antennae. The lines were made by tracing the paths of ants from film. A dot was made every 10 frames, so the closer the dots, the more slowly the ant is moving.

the same place.[37] (This is not just because they are crowded; they turn much more often than they would need to just to keep out of each other's way.) Adjusting path shape to density makes Argentine ants more effective searchers. If you knew that a diamond was lost somewhere in a crowded theater, you might try to organize everyone to look for it around their seat. If there are enough people that someone will search at every seat, someone is bound to find it. But if you had only a few people to search an almost-empty theater, it would be better for each person to cover more ground, searching every row, than for each to look only around her own seat. Ants don't plan out their searching strategies, but the Argentine ants did much the same thing that a plan would dictate: they used straighter paths when there were fewer ants. At low densities, straighter paths are needed to cover more ground. Thus, turning a lot, searching carefully in a particular region, is effective for the group only when there are so many ants that there is likely to be an ant everywhere. When ants are sparse, turning less helps the colony to cover ground more effectively.

A simple algorithm that would produce this behavior is that when an ant meets another, it becomes slightly more likely to turn, with the direction of the turn random. Then the more ants meet each other, the more random their paths. This would lead to more convoluted paths when density is high.

Another species of ant, *Lasius fuliginosus,* regulates contact rate.[38] The ants cluster together when densities are low, and then avoid each other when densities are very high. I learned this by varying density, putting different numbers of ants in different sizes of square arenas. When density increased, with many ants in a small arena, the ants would avoid contact. It seems that these ants can see each other about an ant length away and swerve to avoid meeting and touching antennae.

As density decreased, with fewer ants or a larger arena, the ants were more likely to stick to the edges of the arena. This is a way to keep interaction rate up, because the perimeter of the arena increases only linearly, as the sum of the lengths of all the sides, while the area of the arena increases geometrically, as the product of the sides. However, it might be that *L.*

fuliginosus just likes edges. I tested this, in one of my favorite experiments of my own, by varying the density of ants in a place where they had no edge to prefer—on a sphere. I covered a soccer ball with a nylon stocking, to give the ants some traction, and hung it from the ceiling by a wire that the ants could not climb. I varied the density of ants on the sphere by putting different numbers of ants on the ball. The fewer the ants on the sphere, the more they clustered together. This wasn't just because they like edges—on the sphere, their clusters kept interaction rate high.

Why do ant colonies regulate interaction rate? If interaction rate has an important function, then unregulated interaction rate could lead the colony to react too much. For example, Steve Pacala, Charles Godfray, and I modeled the situation when interaction between a successful forager and an inactive ant stimulates the inactive ant to go out and forage.[39] If the successful ant encounters too many inactive ants, it could stimulate more foraging than the food supply warrants. In general, when contact rate is random, and each ant may contact each of the others, interaction rate changes as the square of the number of ants, so small changes in density could have large effects on contact rate. The active regulation of interaction rate may serve to keep it within an appropriate range. The 11 vehicles in the DARPA Urban Challenge (see chapter 1) had an extreme version of the same problem, because their goal was to avoid contact altogether.

Leaf-cutter ants regulate contact rate as they travel back to the nest from the trees where they cut leaves. When an ant on the way out meets an ant coming back with fragments of leaves to feed the colony's fungus garden, the unladen ant steps aside to let the laden ant pass. Not all of the ants returning to the nest carry leaf fragments. Audrey Dussoutour and others found that the ants coming back empty-handed avoid meeting outgoing ants.[40] They suggest that this saves time on the return trip; by avoiding contact, they also avoid having to step aside.

Interaction rate may be regulated in many ant species. Species probably differ in whether it is important to regulate the rate of interaction, because they differ in the function of interaction. We compared how food influences interaction rate in laboratory colonies of the red imported fire ant *Solenopsis invicta*, *Myrmica rubra*, and *Lasius fuliginosus*.[41] Contact rate seemed to be consistently steady in *L. fuliginosus*, a species that relies on a steady food source, aphids, and to vary most with changing conditions in *Solenopsis invicta*, a species that reacts quickly to make use of disturbed environments. For *Myrmica rubra*, contact rate increased around food, but only inside the nest. In fire ants, encounter rate is higher in the presence of food, either in a familiar setting or when exploring a new region, while in *L. fuliginosus*, contact rate is higher without food (see figure 3.1).

The work of Ken Ross and others on the genetic differences associated with queen number in fire ants suggests another function for interaction networks in this species. Ross and others showed that if enough of the ants in a colony, about 15%, have a certain allele, the colony will accept extra queens into the nest.[42] This could be a response to interaction rate. Perhaps the ants respond to the rate at which they meet other ants that have the polygyne allele, *b*, which seems to affect the odor of the ant that carries it. Perhaps if the rate of interaction with *b* reaches a certain threshold, possibly leading workers to broaden the range of odors they include in their experience of nestmates, then the workers are more likely to accept a foreign queen.

The uses of interaction patterns may be as diverse as the ants. Harvester ants use interaction rates to regulate foraging; the acorn ants *Temnothorax* use them to assess nestmate density in a new nest site; *Pheidole pallidula* uses interaction patterns, mediated by trail pheromone, to get large ants out when needed to retrieve large prey. If I could measure one behavioral trait for every ant species, it would be the rhythm of interactions. We could put them all together to hear the whole symphony of ant diversity, with the percussion from the opportunistic and touchy ants of the tropics, the basso ostinato of the sedate red wood ants as they move up and down the same trees decade after decade, the shifting melodies of the leaf-cutter ants who flow into trees and strip them bare, and the rare grace notes of the ponerines that meet briefly as each one stalks its prey.

4

COLONY SIZE

The behavior of a network depends on its size. The size of an ant colony determines how often ants meet and how many different individuals each ant encounters. Ant species differ greatly in the size at which the colony is mature and ready to reproduce. In all species, a colony starts out small and then grows, beginning with the queen that produces the first workers, and then adding workers until the colony reaches a mature size. How does the size of a colony determine its behavior?

Colony Growth

The Founding Stage

Like any organism, an ant colony has a life cycle (figure 4.1). The most common form of ant colony life cycle begins with a mating flight. Winged reproductives from all the nearby colonies leave the nest at the same time and aggregate somewhere to mate. A new colony is born when a newly mated queen, or a group of queens, establish a nest and begin to lay eggs. The queen spends the rest of her life just laying eggs. During the founding stage, the queen feeds and cares for the brood. When the first eggs grow into larvae, she must feed them. In some species, the queens feed the larvae by regurgitating a metabolized version of their own fat stores, and in others, the queens leave the nest to collect food. Eventually, the larvae become pupae, and then, once the first adult workers emerge from the pupae, these workers begin to enlarge the nest, collect food, and care for the next cohort of larvae. Over time, the colony grows in size. It is mature when it begins to produce winged reproductives of its own. These reproductives mate with the reproductives of other colonies, and the mated females can found new colonies. Colonies die after the queen or queens die once all of the workers have died.

Among ant species, there are many variations on this basic life cycle. Colony life spans range from months to decades, and in some species, a daughter colony replaces its parent colony in the same nest, so that a nest houses a colony that appears to last for centuries. In some species, colonies are founded by a single queen; in others, queens build a new nest together but then fight, and only one survives; in others, a group of queens lives together throughout the life of the colony, sharing the production of eggs. The rate of colony growth, from first eggs to mature size, varies hugely among species, as does mature colony size, ranging from the tiny colonies of some ponerine species, with about 50 workers, to those of many millions, such as the leaf-cutter ants of the New World tropics. There are twists on the basic plan—for example, parasitic species in which a queen goes to live in another species' nest and merely produces daughter queens, all cared for by the workers of the host species; or the slave-making ants, in which the first batch of workers goes to collect pupae of another

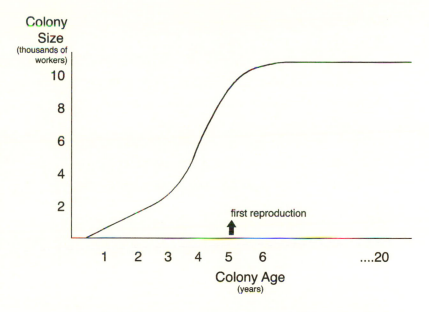

Figure 4.1. Colony growth in harvester ants. A colony is founded by a single queen and grows to a mature size of about 10,000 workers by the time the queen is about 5 years old. This is when the colony begins to reproduce, sending males and daughter queens to the annual mating aggregation.

species and brings them back to join the work force of their colony.

Some of the outstanding questions about colony develop-ment concern the physiological processes that allow a colony to grow: What triggers the development of workers or queens from a female egg, and what leads workers of some species to differ in size? What sets the size at which the colony stops growing? How does an ant queen manage to use sperm from one mating session early in her life to fertilize eggs year after year? Here we leave aside these physiological questions to consider how the colony's behavior develops as the colony grows.

There are not many studies of the development of ant colonies, because ant colonies live a long time, and to track them requires long-term data that follow individually labeled colonies as they grow. It is difficult even to obtain counts of ants in colonies of known age—since counting all the ants kills the colony, every count eliminates the possibility of finding out how much larger that colony would get later on. I've been able to track colony development in harvester ants by censusing the same population for 23 years.[1] Walter Tschinkel was able to measure colony growth in fire ants using a site that was recently cleared so that he knew that all the colonies had been founded since that time.[2] But with so few studies of colony development, we cannot yet provide any general answers to ecological questions such as: How do the first few workers manage to survive? How does colony organization change as the colony grows larger? Does the colony work differently when it is growing from when it reaches its stable adult size?

It's a mystery how any ants, even in a large bustling colony, know what to do, but ants in a viable colony get a great deal of information from each other. It is especially amazing that the first few ants in a founding colony manage to accomplish anything at all. The first workers to emerge are often much smaller than the workers of a mature colony, perhaps because of the limited food they received from the queen as larvae. Those first workers must make their way in a world very different from the one that workers experience once the colony is older. In a large colony, each ant is surrounded by other ants, each probably carrying smells that identify them to task, and each ant moves through a nest with distinct areas for brood, food storage, and pathways to leave the nest and enter it again. But the first workers operate in a much poorer social and architectural environment.

It is clear that not every ant, even in a large colony, is equipped to take care of itself. A harvester ant from a large colony placed alone in a dish with food and water might survive for a day or even for several weeks, but not for the year she might live if she were part of a colony. Whether the ant manages to find food and water might depend on what it was doing when taken out of its home colony. Perhaps ants of some tasks are more likely than others to explore an alien environment. The first few ants in a new colony may not manage to be versatile enough to do everything necessary for the colony to survive.

In harvester ants, the failure rate of newly founded colonies is enormous.[3] Fewer than 10% of the new colonies survive to be a year old. Perhaps this is because there is so little to guide the first few workers to go out, get food, and find their way back to the nest. Small colonies might not have enough ants to get each other going. Consider the Internet: if the numbers of websites were very small, Google's method of searching the web by following links would not work, because often it would run up against 'dead ends,' sites with no links. Similarly, if each ant relies on its rate of interaction to decide what to do, and there are few ants to interact with, then sometimes an ant would get stranded in a location where there was no one else to meet, and the ant would not do anything. Once the number of ants is large enough, then any ant anywhere is likely to meet another one often enough to proceed, most of the time.

We do not know much about the population biology of many ant species, so we do not know whether the failure rate of newly founded colonies is generally as high as it is in harvester ants. Do new colonies die because of a lack of resources in the environment or because the ants are too incompetent to collect enough food? At what point does the network of interactions among ants become sufficiently robust that the colony can survive? To answer these questions, we will need to monitor the new colonies of many different species in long-term studies of ant populations.

As ant species differ greatly in the eventual size of the colony once it is mature and ready to reproduce, species probably differ greatly also in the threshold size that a new colony must reach to have a good chance of surviving. In harvester ants, it seems that if a colony survives to be 2 years old, with about 1,000 workers, it is almost certain to live to be 20 or 25 years old.[4] Perhaps by the time a harvester ant colony is 2, there are enough foragers that one is likely to run into another often enough to make it back to the nest. Two years is a long time—but harvester ant colonies live for 25 years or more, and they are subject to long-term pressures, such as cycles of about 10 years of drought and heavy rainfall, and competition with neighboring colonies that may live next door for 20 years. In contrast, fire ant colonies live for about 5 to 8 years, less than a third of a harvester ant colony lifetime. In this short time, the fire ant colony grows to contain hundreds of thousands of workers, about 25 times the size of a large harvester ant colony. Perhaps colony survival in fire ants is more tightly coupled to short-term fluctuations in food supply than in the slower, steadier career of a harvester ant colony, and fire ant colonies must get very large quickly to get the food necessary to survive.

Growth: From Adolescence to Maturity

Colonies grow in numbers of workers when worker births outpace worker deaths. This could happen because the production of workers accelerates, or workers live longer, or both. Figure 4.1 shows how the number of workers changes over time in a harvester ant colony. This curve was made from data we obtained by digging up colonies of known age and counting all the ants.[5] The sigmoidal shape of the curve, with rapid growth leading to a stable, mature size, is thought to be characteristic of ants, as Oster and Wilson suggested in their 1978 book. However, since we know little about the growth of colonies in natural populations of most ant species, we do not yet know how typical this curve is.

In harvester ants, fire ants, and perhaps many other species, once a colony moves past the founding stage, it enters into a period of rapid growth. As more workers are produced, they can get more food, which makes it possible to produce more workers. Much more food is needed to produce an ant than to produce an egg, because ants are much larger than eggs. As Tschinkel points out in *The Fire Ants*, an adult fire ant female reproductive is 3,500 times as large as an egg. Food probably limits not the number of eggs a queen can lay, but the number of eggs that survive to become workers. In many species, some eggs are consumed, going back into the colony food supply. Presumably, this occurs when food supplies are low.

We don't know exactly what sets the rate of colony growth in any natural population of ant colonies. To understand how the number of workers produced is regulated in response to food supply, we'd need to know what the ants eat and how that food fluctuates in availability, whether the colony stores food or has to use food as soon as it is obtained, how quickly eggs turn into larvae and larvae become workers, and what system is used to distribute food to larvae. Tschinkel's work on fire ants addresses some of these questions. Fire ants regulate the number of new workers produced, using temperature, in relation to food supply. The warmer it is and the more food available, the faster the colony grows. Debby Cassill's work shows that each time a larva is fed, it gets the same amount of food, so when there are more larvae, more workers are required to feed them.[6] Colony growth stimulates further growth, because the more the colony grows, the more workers are available, both to get more food and to feed it to more larvae.

Army ants are nomadic, and they eat as they move. They range in body size from the large *Eciton* in Africa, who sweep across the countryside in waves, forcing people to put the legs of their beds in buckets of kerosene to keep the ants out, to *Neivamyrmex*, a scourge for other ants but too tiny to be much noticed by people. During the raiding phase, army ants collect insects and the brood of other ants. Then they find a temporary nest, or bivouac. In one of the first demonstrations of collective behavior in animals, T. C. Schneirla's work in the 1950s showed that the timing of raids is related to the presence of brood.[7] During the bivouac phase, the queen lays eggs. Then when the eggs become larvae, the larvae must be fed, so the colony goes on the march again to collect food. This was the first clear example of how a colony's growth drives its relation with its environment. Other insects follow along the army ant trail and do their own foraging on prey dropped by the ants.

For ant colonies that live in mutualistic relations with plants, growth has unique consequences. The plants provide nesting space for the ants. Often, they also feed the ants, either directly through nectar bodies or indirectly by providing resources for scale insects that suck the phloem of the plant and excrete a sugary solution that feeds the ants. In turn, the ants defend the plant from herbivores. By protecting the host plant, the ants promote its growth. Larger colonies provide more effective defense, so as the colony grows, it helps the plant to grow more nest sites. This in turn allows the colony to grow more, because a larger plant provides both more nesting space for the ants and more resources for the scale insects that feed the ants. Thus, as Megan Frederickson has shown in studies of Amazonian ants and their host plants,[8] the growth of a colony fuels further growth. This is not only because, as in any ant species, a larger colony has more ants to obtain more resources, but also because a larger colony actually produces more resources for itself, by promoting the growth of its host plant.

Task Allocation and Colony Size

As colonies grow, there is a shift in the allocation of workers to various tasks. There are data on this for only a few species, but these all show that a higher proportion of the colony is devoted to foraging when the colony is young and small than when it is older and larger. The result is that the total number of foragers does not increase linearly with colony size. For example, counts made of foragers outside the nest show that in a young harvester ant colony with 1,000 ants, about 500, or 50%, are foragers; in an old colony with 10,000 ants, about 2,000, or 20%, are foragers. The 4-fold increase in number of foragers, from 500 to 2,000, is much smaller than the 10-fold increase from 1,000 to 10,000 in total colony size.[9]

In *Lasius niger*, as in harvester ants, numbers of foragers grow more slowly than total colony size. Claire Detrain and colleagues manipulated the growth of *L. niger* colonies in the laboratory by adding brood to nests with newly mated queens.[10] They compared the patrolling and foraging behavior of smaller colonies that grew without intervention and larger colonies to which extra brood was added, observing the behavior of large and small colonies after the added brood had become adults. The numbers of patrollers, and the numbers of foragers that showed up immediately after food was provided, both increased with colony size, but did not scale linearly with colony size. There were only about 0.2 more patrollers for every new ant in the whole colony, and about 0.4 more foragers for every new ant in the whole colony.

Tschinkel's studies of the growth of fire ant colonies indicate the same trend. However, changes of allocation to foraging in growing fire ant colonies cannot be easily estimated because the measure of colony size used in Tschinkel's study, an extrapolation from the number of ants in a known amount of soil of an excavated nest, did not take into account the foragers who were outside the nest when it was excavated.

As an ant colony grows, if all the new ants are not foraging, what are they doing? Surprisingly, the answer seems to be that they are doing nothing. As a colony grows larger, the proportion of reserves grows. In harvester ants, 50% of a 1-year-old colony forages, while only about 35% of a mature colony forages. In laboratory colonies, it looks like the numbers tending brood do not increase linearly with the amount of larvae; when the pile of brood is large, there are not many more ants working there than when the pile is small. So if the proportion foraging decreases as the colony grows, and the proportion working inside the nest doesn't increase with colony size, the proportion doing nothing must increase.

It seems that, despite the use of ants in Proverbs ("Look to the ants, thou sluggard . . . ") and elsewhere as role models for the work ethic, a lot of ants are just hanging around. Perhaps they are needed as reserves, available to be called out to work if they are needed, although in my 25 summers of watching harvester ants, such an emergency has never happened. Twenty-five years is a long time for a human observer, but not even close to a blink in evolutionary time. Another possibility is that these inactive ants are a means of food storage. Honeypot ants (*Myrmecocystus*) make use of ants for food storage in a spectacular way. An ant may become a replete, an ant that stores so much liquid food that its abdomen swells up to the size of a pea, which obliges the ant to cling to the ceiling, unable to walk around since its little legs cannot reach the ground past its enormous abdomen. Repletes regurgitate their stored nectar to other ants. In other species, some ants could be receptacles for stored food, even without such dramatic changes in appearance.

Another possible function for lazy ants is that a buffer zone of inert ants may help to dampen the interaction rate or the response to interactions. The function of this buffer zone may be analogous to what some think that sleep does in our brains: while we sleep, the connections most heavily reinforced by the day's events persist, while all the incidental connections are discarded. The inert workers may absorb interactions, so that only the ones repeated most often have an effect on the colony's behavior.

Task fidelity, how likely an individual is to keep doing the same task, changes as the colony grows larger. In a harvester ant colony, the ants in younger colonies switch tasks more readily than the ants in older ones. Ants marked one day as nest maintenance workers might show up the next day as patrollers or foragers. In an older colony, ants tend to perform the same task from one day to the next unless there is a drastic change in the environment. If no new food becomes available, if there is no big mess to clean up outside the nest, if there is no incursion by unfriendly neighbors—then most ants will do the same task they did the day before. This difference between young and old colonies could be a consequence of limited numbers of ants. Perhaps in a small colony, when more ants are needed to perform a certain task, the extra ants have to come from those doing some other task, while in a larger colony there are ants just hanging around who can step up. For example, in the case of foraging, we know that the stimulus to forage is the return of successful foragers. In a small colony, if there are not many foragers waiting around to go out, the returning foragers probably contact workers of another task group, and those are eventually stimulated to forage themselves.

Task fidelity depends on colony size even in the small colonies of the ponerine ant *Rhytidoponera metallica*.[11] Colonies grow to have about 450 workers. Melissa Thomas and Mark Elgar found that in smaller colonies, just as in harvester ants, each worker was likely to perform more different tasks. This depended on the size of the colony but not on the age of the worker. Young and old workers both performed about the same number of tasks, although younger workers spent more time grooming themselves. Workers in large colonies spent more time foraging than workers in small ones, regardless of the age of the individual worker. These results are consistent with a system in which interaction increases the probability of continuing a task, but this has not been studied directly. In larger colonies, interaction rate is more frequent and more steady than in smaller ones. In a small colony, when contact is more infrequent and more variable, workers are more likely to switch tasks.

Not only the task fidelity of individuals, but also the responses of colonies change as harvester ant colonies grow older and larger. Task allocation in older colonies is more consistent, and more homeostatic, than in younger ones.[12] It seems that younger colonies are more sensitive than older ones to whatever changes in the environment from one week to the next, such as the weather or food availability. I did a series of perturbation experiments with large colonies, more than 5 years old, with about 10,000 ants. I did the same experiments with young colonies, 2 years old, with about 1,500 ants. I repeated the experiments week after week. Older colonies tended to respond the same way each time to a given perturbation. But the younger colonies responded differently each week. However, the variation among colonies in a given week was about the same in older and younger colonies.

The second difference between young and old, or small and large colonies, was that larger colonies behaved more homeostatically, returning to the baseline or undisturbed state as conditions worsened. This difference showed up when I increased the magnitude of perturbations and interfered with more than one activity at once. For older colonies, there was a nonadditive effect. If I caused more nest maintenance work by putting out piles of toothpicks that the nest maintenance workers cleaned up, then the colony did less foraging. If I decreased foraging, by putting out little plastic barriers that partially obstructed the foraging trails, then the colony did more nest maintenance. But if I performed both experiments at once, the colonies did about as much foraging as undisturbed colonies. Somehow when there are multiple disturbances, large colonies respond less to the disturbance and concentrate on foraging.

Individual harvester ants live about a year, so the ants in an older colony are not any older than the ants in a younger one. The differences in behavior of older and younger colonies cannot be attributed to the accumulated experience of individual ants. The simplest explanation is that the ants always respond to conditions according to the same algorithm, but an ant experiences different conditions in a large colony than in a small one. This is what led me to investigate the role of interaction rates in the first place. My reasoning was that an ant in a small colony is likely to interact with other ants at a lower rate than an ant in a large one. So if an ant uses the rate of interaction with others to decide what to do, then the behavior of old, large colonies will differ from that of young, small ones. That is, the ant in an older colony can follow the same algorithm in responding to interaction as an ant in a younger colony, but the outcome will be different in a larger colony because its interaction rate is different.[13] Now we know that in fact, ants do use the rate of interaction with others to decide what to do.

A larger colony has more ants to participate in its network of interactions. Steve Pacala, Charles Godfray, and I modeled one of the consequences of this.[14] The larger the colony, the more likely it is that when something happens in the

environment, some ant will detect it. If ants transmit information when they meet, the superior powers of detection of
a large colony will make it respond more quickly than a small
colony to its environment.

In general, as Lauren Meyers pointed out to me, the stabil-
ity of larger colonies could be due to the improved sampling
that large size provides. Each ant's interactions with others
is a sample of some aspect of the current state of the colony,
such as the rate at which foragers are coming in with food,
or the relative numbers of foragers and nest maintenance
workers currently active. The larger the colony, the more
accurately each ant will measure the true state. Or to put it
more precisely, the smaller the colony, the more opportunities
for sampling error to affect an ant's interaction experience. When there are few ants, each ant is more likely to
interact with only a few ants, and those few might not be
representative.

Interaction networks probably change in many ways as a
colony grows older and larger. For example, larger colonies
have larger, more elaborate nests. Nest architecture determines
how ants and resources are distributed, which influences the
rate at which traffic flows, and these in turn influence the
rate of interaction among ants. The underground nest of a
harvester ant colony has a more complicated tunnel system,
with more than one chamber just inside the nest entrance,
than a younger one. Species differ in how the nest grows as
the colony does. In some species, it seems that the nest of a
larger colony is larger than that of a young one, but it keeps
the same form—for example, extending a central trunk and
adding more chambers that branch off it. As colonies grow,
changes in the nest, and in the relations between nests in

species that have more than one nest for each colony, produce changes in interaction patterns.

Ecology, Behavior, and Mature Colony Size

The ecology of an ant species is closely linked to the size of its mature colonies. Many species of ponerine ants, especially in the tropics, have tiny colonies, some with less than 50 ants. Ants in small colonies seem to function more independently than ants in large ones. An ant's behavior arises from the combination of its responses to environmental cues and interactions with other ants. Perhaps the smaller the colony, the more each ant has to rely on environmental cues, and the larger the colony, the more it can rely on interactions with others.

Ray Mendez, a zoo designer and insect photographer, once told me a story that illustrates this. He was searching for the nest of a *Paraponera* colony in the jungle in Thailand. The usual method for finding nests of any ant species is to put out bait, wait until an ant picks it up, and follow the ant back to the nest.[15] *Paraponera* are enormous ants with small colonies and a very intense sting. They are carnivorous (although I met some in the jungle in Mexico that were very fond of chocolate cake, but that's another story). Ray found a cricket, killed it, carefully put it down in the path of a *Paraponera* worker, and sat down to wait. The ant came along and picked up the cricket. Ray held his breath waiting to see where the ant would go. The ant ate the cricket and walked on. This shockingly self-centered behavior is probably unique to species with very small colonies, and reminiscent of the behavior of the ants' ancestors, the solitary wasps. It may be that when an ant colony is small enough, each ant feeds itself and the brood, but does not contribute to any centralized store of food.

Working together to obtain food is rare and clumsy in the ponerine ant *Gnamptogenys sulcata*, a species with tiny colonies with only a few foragers, common in cocoa and coffee plantations in Mexico. The ants are predators of insects. S. Daly-Schveitzer and others found that if a forager finds an insect up to about 8 times its own weight, it kills the prey and brings it back to the nest alone.[16] If it finds a larger insect, about 20 times its weight, it goes back to the nest and rouses the other foragers to come back to the prey. But the recruited foragers made it to the prey in only 60% of cases. When they did reach the prey, they all dragged it back toward the nest in bouts. When prey was pinned to the ground and so could not be moved, it attracted many more ants, apparently because the ants perceived it as heavier and thus more worthwhile, although the ants were unable to get any of it back to the nest.

Differences among individuals could be more important when colonies are small. If there are 10 ants in the colony, and all are lazy, the consequences for the colony will be worse than if a colony with 100 ants has 10 lazy ants but has another 90 to make up for them. It is interesting that it is in the species whose mature colonies are still very small that individual variation seems to be most extreme. What has been measured is mostly variation in activity level—some ants are more active than others. It may be that this seems more common in species with small mature colonies, such as the ponerines, only because individual variation has been studied more in such species. But it might be instead that there is selection for more variation among individuals in these species because in smaller colonies, there have to be some especially eager ants to make up for the slackers.

Individual differences in odor also might have more effect on colony behavior in small colonies. As you saw in chapter 3, harvester ants change odor as they shift tasks; ants that work outside as foragers have a higher proportion of n-alkanes in their cuticular hydrocarbons. If an ant happens to meet a new forager that doesn't yet smell much like one, the interaction might not register or might be less effective than an encounter with an ant with full-fledged forager odor. In a young or small colony, such an interaction might be one of few that an ant has with foragers, so an inactive forager in a small colony might not meet enough convincingly forager-like ants to go out and forage. Overall, this effect would diminish the amount of foraging that a small colony does. In a large colony, an interaction with an ineffective forager is likely to be followed by many others, so the chances are larger that an inactive forager will meet enough ants with forager-like odor that it will go out to forage. Overall, this feedback would increase the amount of foraging that a large colony does.

Large colony size brings its own problems. It may be more difficult for a large than a small colony to distribute substances. When the colony is large enough that every ant can't reach every other one, flow must be channeled. One substance that must move around the colony is the colony odor. Every ant's cuticular hydrocarbons carry the colony-specific odor. This is manufactured by each ant, but is also passed around when ants groom each other. Initially, the odor that is put on by other ants is more important than whatever the ant secretes itself. We know this from the experiments of Adele Fielde in the early 1900s.[17] In many combinations of species, she introduced brood—larvae and pupae—from one species into a colony of another and found that it was accepted. It seems that brood do not carry a colony-specific odor, or they are so frequently groomed by the workers that they quickly acquire the cuticular hydrocarbons of the adults. The behavior of slave-making ants also shows that colony odor must be passed around through interaction and grooming. They capture brood from another species and bring it back to the captors' nest, and there the workers live out their lives, doing the work of the colony. Again, the newcomers' foreign odor is either superseded by that of the slave-makers or ignored.

Cuticular hydrocarbon profiles change over time and have to be renewed. For example, Philip Newey and others found that in the weaver ant *Oecophylla smaragdina*, the profiles of intact colonies and of ants taken out of the colony both changed in parallel over the course of about 6 weeks.[18] As food supply and other unknown factors change an ant's hydrocarbon profile, the colony odor of adult workers must be renewed frequently. The larger the colony, the more complicated the distribution of colony odor.

In very large colonies, the numbers of ants engaged in a task could be so large that they are unlikely to be able to communicate. Leaf-cutter ant colonies can have many millions of ants in an enormous nest; one famous photo of an excavated nest shows a hole that dwarfs a backhoe and many people standing inside it. In a very large colony, the organization of the colony must allow for interaction within subsets of workers. The larger the colony, the more likely there are to be disparities in the sizes of groups that interact, such as in the numbers of workers that bring in food and that process it inside the nest. The models of Francis Ratnieks and Carl

Anderson suggest that an effective way to deal with such disparities is when the response of each individual in the smaller group requires multiple interactions with individuals in the larger one.[19] In extremely large colonies, such patterns of interaction may be needed to accomplish the distribution of colony odor and of food, to ensure that food gets to the larvae and that information about the current needs of the brood gets to the foragers.

In species in which one colony occupies many nests, there are many opportunities for the size of a functional nest unit to change. As ants flow between nests, occupy new ones, and abandon others, the colony organization must adjust to shifts in colony size. Joan Herbers and colleagues have studied species that move from one small cavity in a twig or an acorn to another, such as some *Leptothorax* species or *Myrmica punctiventris*.[20] Shifts from one nest to another go along with seasonal changes in colony size; perhaps moving around among nests is a way to achieve particular colony sizes. The invasive Argentine ant has many nests, and nest size changes seasonally; a colony coalesces into a few large nests in the winter and disperses into many small ones in the summer.[21] Many species of red wood ants in northern Europe have huge colonies with many large permanent nests and carry brood and ants from one nest to another. In all of these arrangements, colonies keep changing how many ants are working together and how each nest is connected to the others of the same colony.

The intriguing questions about ant colony size and growth are questions about the effects of network size in general. Very small organizations work differently from larger ones.

How much of this is due to variation among the participants, which would have more effect when the organization is small? How much is due to the pattern of interactions, which has a very different structure in a larger organization? At what point does a system become so large that it must operate in subsets? Are animals with larger brains always smarter, and why?

5

RELATIONS WITH NEIGHBORS

Scaling up, we can consider the relations among colonies as a network of interactions. Colonies interact directly when ants of each colony meet, and they interact indirectly when one colony uses a resource that the other also might use. The outcome of interactions between colonies depends on the same parameters as for ants within colonies: how often the colonies meet, what happens when they interact, and how long the effect of the interaction lasts.

To consider relations between colonies, we have to add a level to those of ants and colonies: populations of colonies. 'Population' is a technical term in ecology that means the set of individuals of a species that could reproduce with each other. For ants, what is an individual? Although ants are the individuals we see walking around, they are not individuals in the ecological sense, because ants don't make more ants; most of the ants in a colony are sterile workers. Instead, it is colonies that make more colonies. So colonies are the reproductive individuals, and a colony is not a population of ants. An ant population is all of the colonies that are close enough that their reproductives might mate with each other.

The relations among ant colonies include neighbors of the same and of different species. We know relatively little about how colonies ineract. The paucity of studies on what happens when colonies meet is mostly due to the boundaries that distinguish fields of biology. Until recently, ecology and behavior were distinct fields, studied by different people. Ecologists thought about how organisms use resources and how this explains the numbers and distribution of species, and so ecologists identified and counted ants. Behavior was about what animals do—for example, what happens when animals meet—so people who studied animal behavior watched ants to learn how ants get things done. But this distinction is misleading; behavior and ecology cannot be separated. How many of a certain species live in a certain place, its ecology, depends on what that organism does to get resources, its behavior. Now the divide between behavior and ecology is gradually fading away. People are beginning to study both what ants do to work things out with their neighbors and also how the outcome of interactions produces the numbers and distributions of each species in the community.

Relations with Neighbors of the Same Species

Interactions of neighboring harvester ant colonies shape their foraging behavior. Colonies compete for food with their neighbors of the same species. Harvester ants forage for seeds that are distributed on the ground by wind and flooding, so no matter what plants happen to grow nearby, those plants don't provide reliable food for the ants.[1] Instead, the food in one place is just as good as the food in any other. In the long term, the amount of food a colony gets depends mostly on how many ants go out to search for it, and not on where they go. How often and how much a colony foragers matters to its neighbors, because the more thoroughly one colony searches an area for food, the less food is left for its neighbors to find.

Competition with neighbors for food is an important pressure on harvester ant colonies. Large neighbors of the same species diminish a new, small colony's chance of survival because the large neighbors take so much of the available food. I have censused a population of about 300 colonies every year since 1985. Every year, we map the locations of all colonies, add the new colonies to the map and remove the ones that died.[2] Looking at these maps, year after year, it turns out that young colonies with large neighbors are less likely to survive.[3] Another line of evidence that food is scarce is that after drier years, when plants produce fewer seeds, colony mortality is high.[4]

Competition for food has shaped how the foraging behavior of one colony responds to that of its neighbors. Colonies interact frequently with their neighbors. When the trails of neighboring colonies meet, the foragers of neighboring colonies are searching in the same area, and so they are competing for food—what one colony picks up, the other will not get. One summer, we mapped the trails of 34 colonies for 17 days. Each colony uses many trails per day, an average of about 4. We found that on average, 1.9 trails per colony per day met a trail of a neighbor.[5]

If the foraging trails of neighboring colonies meet one day, both colonies are likely to use different trails the next so that they don't meet again in the same place.[6] Avoiding overlap with neighbors seems to be an important function of the patrolling system. It appears that an encounter with a neighbor makes a patroller less likely to reinforce a foraging direction. Patrollers set the direction of the day's trails, by putting down a chemical cue on the nest mound.[7] The patrollers move around the foraging area early in the morning before the foragers are out. Sometimes the patrollers of neighboring colonies meet. Here is a scenario that is consistent with our observations. The patrollers explore the directions used the previous day. The day after two foraging trails met, the patrollers of each colony will return to the previous day's foraging area, making it likely that they meet. Each patroller that met a neighbor's patroller avoids returning directly to the nest. Coming back to the nest from another side, it doesn't put down its chemical secretion in the direction that led to the encounter with the neighbor. Meanwhile, other patrollers that went in other directions and did not meet the neighbor's patrollers return to the nest and do put down the secretion that leads foragers back the way they came. This is sufficient, most of the time, to lead foragers away from the site of the encounter with the neighbor's trail, and since both colonies do this, the two neighbors are unlikely to meet in that place two days in a row.

However, it is only mature colonies, older than 5 and at their stable size, that avoid foraging toward the place they met a neighbor the previous day. Colonies that are only 3 to 4 years old, not yet reproductively active and at the steep part of their growth curve (see figure 4.1), return day after day to the site of where foragers of the two colonies overlapped (see figure 5.1).[8] This adolescent belligerence may be the result of growth rate. A mature colony has to produce 10,000 ants to maintain its size from year to year, but it has 10,000 ants to do the work necessary to produce the next 10,000. A 3-year-old colony has only 6,000 ants to produce the 8,000 ants in a 4-year-old colony. Thus, the demand for food, per forager, is greater in the smaller, growing colony than in the larger one of stable size. Fred Adler and I made an optimality model that suggests that this would be a good reason for the ants to act as they do,[9] but to find out what actually produces the behavior of adolescent harvester ant colonies, we would need to learn more about processes that occur inside the nest. We do not know how the larvae's demand for food is transmitted to the foragers. Perhaps the demand for food influences the patrollers' decisions, so in younger colonies the patrollers are less likely to avoid a certain direction where they met the neighbors' patrollers.

Although the exceptions are spectacular, ants don't often fight with each other. In many species, ants of neighboring colonies avoid each other, and an encounter between two ants is enough to persuade both of them to go off in different directions. Ants sometimes look like they jump apart after an encounter with an ant that is not a nestmate, recoiling from the unfamiliar smell. Sometimes there is extensive mingling of ants from different colonies without any aggressive action. In general, how ants react to each other depends on the situation. One way to test whether ants belong to the same colony, or whether they act aggressively toward ants that are not nestmates, is to put some ants in a dish to see if they will fight. But we rarely know whether the conditions for outright violence are met, and this is why the outcome of such experiments is so variable, as Tai Roulston and colleagues showed.[10]

Fighting is rare in harvester ants. Harvester ants have very toxic venom; perhaps this has encouraged the evolution of avoidance. There seem to be seasonal bursts of fighting, often

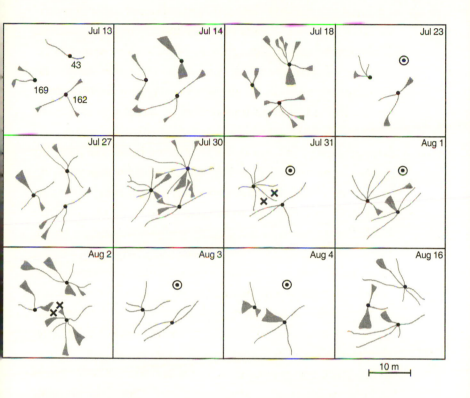

Figure 5.1. The interactions of three young harvester ant colonies. The colonies, all 3 or 4 years old, are named '169,' '43,' and '162.' Older colonies avoid foraging in the directions where they met their neighbors, but the trails of young neighboring colonies, like these, meet over and over. Although all the trails overlapped on 30 July, the colonies continued to meet over the next two weeks. Nest entrances are shown as solid circles, the black shapes show where the colony foraged that day, and a solid circle with another circle around it means that colony (always colony 43 in this figure) was not active that day. 'X' shows the location of fighting.

just after the summer rains. Maybe the rain washes away chemical signals on the ground, such as colony-specific cuticular hydrocarbons, and the absence of those signals stimulates fighting. When harvester ants do fight, one ant clamps its mandibles onto the other's petiole, the segment between thorax and abdomen. The attacker tries to cut the other one in pieces. Often it does not succeed and dies of desiccation, still clamped on. Eventually, its body falls off but its jaw muscles remain contracted, even in death, so the attacked ant spends the rest of its life walking around with its attacker's head attached like a little trophy on its belt.

We know that ants recognize nestmates through cuticular hydrocarbon profiles, but we do not know exactly how. One idea is that an ant uses its own odor as a template and responds to the match between its own odor and that of an ant it meets. We also do not know how much of the variation among cuticular hydrocarbon profiles depends on genetic relatedness. Norm Carlin's experiments with carpenter ants,[11] like Les Greenberg's pioneering ones with sweat bees,[12] suggest that there is less hostility between colonies founded by closely related queens than those founded by unrelated queens. Cuticular hydrocarbon profiles also depend on food intake, as shown by the hostile reaction of Argentine ants to others fed a certain cockroach (see chapter 3).

Nearby colonies might be similar in profiles because they use similar food. Yet an intriguing study of fire ants by Robert Vander Meer and others showed the opposite: the closer the neighbors, the more they differ in hydrocarbon profile.[13] This suggests that somehow colonies modify their hydrocarbon profiles in reaction to each other, but how this works remains anyone's guess.

The reaction of ants to their neighbors of the same species seems to depend on the rate at which they interact, not just how much they differ in odor. Ants can distinguish the odors of neighbors from those of more distant colonies. In harvester ants, the foragers react more to the neighbors, within range of their foraging trails, than to ants from more distant colonies. Contact with ants of a neighboring colony of the same species inhibits foraging more than contact with ants far away.[14]

How do ants learn the odors of the neighbors? There are many hundreds of harvester ant foragers in a foraging trail, and it must be rare for each one to meet a neighbor. Using counts of numbers of foragers and the probability that neighboring colonies meet, and estimating from our observations of marked ants that an ant is a forager for about 30 days, it seems that each day any forager has a probability of only about 0.06 of meeting an ant of another colony, and a lifetime probability of interacting with only about 2 ants of neighboring colonies. Mark Brown and I wondered if certain ants specialize in recognizing neighbors.[15] We kept two colonies in a laboratory arena, separated by a sliding door. We opened the door and marked the ants that went through to visit the arena of the other colony. We did not find that there were any ants specialized in diplomacy; there were many encounters, but not between the same ants.

Since each ant meets one of its neighbors rarely (although colonies meet often), and there are no special ant diplomats, neighbor recognition must develop very quickly and persist over many days. Mamiko Ozaki and others found that in the carpenter ant *Camponotus japonicus*, the antennae respond only to the cuticular hydrocarbon blends of non-nestmates, not to those of nestmates.[16] It seems that the ants become so familiar with their own colony's odor that only the odor, of a non-nestmate elicits a response. If this occurs in harvester ants, it explains why they react to neighbors with little prior experience of the neighbor's odor, but does not explain how they distinguish neighbors from strangers.

While harvester ants act more concerned about neighbors than ants from far away, some ants do the opposite. In a study of two desert species of seed-eating ants, *Pheidole tucsonia* and *P. gilvescens*, Peter Nonacs and others found that colonies were more hostile to distant colonies than to near ones.[17] This was true in encounters between colonies of different species as well as with colonies of the same species.

Perhaps the extent of hostility between neighboring colonies depends on whether they meet often enough to recognize each other, however rapidly such recognition develops. If there is no neighbor recognition, response may depend only on how much their hydrocarbons differ as a result of differences in food supply. For example, colonies of the meat ant *Iridomyrmex purpureus* consist of several nests, often hundreds of meters apart and connected by trails. Ellen van Wilgenburg, extending the work of Melissa Thomas and others, showed that ants from adjacent colonies are likely to fight, and this is independent of genetic relatedness.[18] However, within a neighborhood, ants fight more with more distant colonies, perhaps because differences in food supply lead to differences in odor. Results from several studies of neighbor recognition in this species suggest that aggression is more likely when colonies are crowded and more likely to meet.

Some ant species are never hostile to ants of another colony. Ellen van Wilgenburg and others in Mark Elgar's lab found that in the Australian bulldog ant *Myrmecia nigriceps*, an ant can walk into the nest of another colony, and no one reacts.[19] This does not seem to be a friendly relationship that develops over time, because there is no difference in the reaction to neighbors or ants from far away. It seems that these ants simply don't distinguish nestmates from ants of another colony.

The most extreme cases of friendly relations among nests occur in unicolonial species, in which it is difficult to say where one colony ends and the other begins, because there are many nests linked by trails, with ants traveling from one nest to another. Over many years, Daniel Cherix recorded the paths among nests of the red wood ant *Formica rufa* in a reserve in the Swiss Alps.[20] These ants carry brood from one nest to another. Cherix traced one network of paths covering more than 12 hectares, an area 1,200 meters long on one side, that connected 27 large nests and many more small ones. The pioneering studies of population genetics in ants were carried out by Pekka Pamilo and others on the same species group, red wood ants, in Finland.[21] They found that the ants within a nest were not any more closely related than ants of different nests. Mixing among nests, as ants move from one nest to another, contributes to this, along with inbreeding and mating between reproductives of nearby nests.

Many studies show that the outcome of conflict between neighboring ant colonies depends on the relative size of each colony. Eldredge Adams did an experiment with *Azteca trigona* in tropical forest in Costa Rica.[22] He set up a platform near a foraging trail of one colony and then connected it by a small bridge to another platform near the trail of a second colony. This led the two colonies to fight. Adams considered the colony that occupied both platforms once the fighting was over to be the winner; if both colonies were still on the platforms, it was a draw. The colonies varied in size from 14 to 300 workers. Usually, but not always, if the colonies differed in size, the larger colony won. Priority, who got to the platform first, as well as numbers, seemed to influence the outcome. What decided priority was how quickly each colony recruited nestmates to the fight. In conflicts between two colonies of equal size, when one colony occupied the platform first, the colony that was there first always won.

Just as ants within a colony use interaction rate as a cue to the numbers of nestmates performing a task, or finding a nest or food source, so in interactions between colonies, ants can use interaction rate to assess the numbers of ants from another colony they are dealing with. The rate at which ants of one colony meet ants of the other depends on the relative sizes of the two colonies—that is, the ratio of numbers of workers in one colony to numbers in another.

The use of interaction rates would explain why relative size determines the outcome of encounters between colonies. If ants are using interaction rate to assess which colony is larger, they should respond to the *ratio* of 'them' to 'us,' rather than to the *numbers* of 'them.' We tested whether ants respond to ratio or number of non-nestmates in the ant *Lasius fuliginosus*.[23] We set up groups of host ants in two sizes, either groups of 35 or groups of 75 ants. Then we introduced 15 ants of another colony. In both cases, the ants ran around, alarmed. The smaller group of 35 host ants responded more intensely to the 15 introduced ants than the larger group of 75. An ant's response depended on the rate at which it met an ant from another colony relative to the rate at which it met nestmates. If the host ants' reaction depended simply on the number of intruders, then the reaction should have been the same regardless of host group size, because in both cases, there were the same number, 15 intruders. To my knowledge, this has not been tested in other species, but I think it is likely that an ant's response to an encounter with other colonies depends on the rate at which it meets non-nestmates relative to nestmates.

Colonies of the same species interact indirectly, through their use of resources. For example, harvester ant colonies sometimes move into the nests of their dead neighbors, and this seems to occur more often in years when resources are abundant, in which new colonies flourish and nests are more scarce. The nests are elaborate structures with adobe-lined walls, so it is not surprising that an abondoned fully developed nest of a mature colony is valuable real estate.

Interactions between Species

Ant colonies are always engaged in interaction with neighbors of other ant species. They go about their business near each other, sometimes meeting, sometimes not, and even avoidance is a kind of interaction. Conflict has been studied much more than peaceful coexistence, perhaps because it is more conspicuous, although not necessarily more interesting.

In conflicts between species, numbers of ants seem to matter much more than differences in body size. Although in a fight between two ants that differ in size, the larger ant might win, in a fight between two colonies, the one with more ants wins regardless of size. Often many small ants will grab the legs of a larger ant; if there is a small ant attacking each of the larger one's legs, the larger one is probably doomed.

Considering that most of an ant colony's food goes to turn eggs into workers, and the larger the worker, the more food it requires, it has been argued, by Nigel Franks and others,[24] that the smaller the ants, the more workers in the colony. Of course, there are some species, such as leaf-cutter ants, that produce huge colonies with large ants. In general, however, the argument runs that species with small ants will actually be at an advantage in conflict with other species, simply because there are more ants to show up.

The invasive Argentine ant acts more aggressively toward other species when its colonies are large. Katayo Sagata and Philip Lester created laboratory colonies of different sizes of Argentine ants and of the native New Zealand ant, *Monomorium antarcticum*.[25] The Argentine ants, but not *M. antarcticum*, were less aggressive, and more likely to avoid interaction with the other species, when their colonies were small, but plunged into all-out conflict when there were 1,000 workers or more. In the field, small colonies were as likely to persist as large ones. Adjusting its behavior to colony size seems to help the Argentine ant survive in new areas.

For harvester ants, the most troublesome neighbor is *Aphaenogaster cockerelli*. These ants forage at night. Before dawn, when the harvester ants are still inactive inside the nest, the *Aphaenogaster* plug up the harvester ant nests with pebbles and soil.[26] When a harvester ant nest is plugged, the foragers emerge later, but they have to go back inside when it gets hot at midday, no matter how late they came out. That means that a delayed start shortens the foraging period, leaving more food for that night, when it's the *Aphaeonogaster*'s turn to forage. The *Aphaenogaster* pick on the younger, smaller harvester ant colonies to torment this way. It's a mystery how the *Aphaenogaster* figure out that a nest is young and thus a suitable victim; perhaps in the hour or two their foraging activity overlaps with that of *P. barbatus* on a cool morning, they use interaction rate to evaluate the size of the neighboring colony. Nathan Sanders and I found that from the harvester ant perspective, it is definitely risky for a young colony to live near an *Aphaenogaster* colony.[27] These conflicts can go on for years. I have seen a couple of extended wars between *Aphaenogaster* and harvester ant neighbors. Sometimes the harvester ants move into abandoned *Aphaenogaster* nests, and sometimes *Aphaenogaster* wipe out the harvester ant colonies and move in.

A colony of one species may use the interaction network of the other. Eldredge Adams found that *Cephalotes* ants in mangrove swamps use the pheromone trails of another species, *Azteca trigona*.[28] The *Azteca* find food and recruit rapidly. The *Cephalotes* are slow and plodding. The *Azteca* tolerate the presence of *Cephalotes* at their food sources, and the *Cephalotes* are not likely to arrive in large numbers before the food source can be used by the *Azteca*. Sometimes the outcome of shared pheromone trails is not so happy. R. B. Swain found that the eavesdropping *Camponotus femoratus*, who lives in the same nest with colonies of *Crematogaster limata*, follows the *Crematogaster* trails, and if it gets there first, keeps the *Crematogaster* from the food source.[29]

Many ecological studies of ant communities investigate which species are present and in what numbers, but not how the ants of different species interact with each other. Such studies point indirectly to some insights about behavior, because which species are there, and how many, is the result of how they interact or avoid each other. In the long term, many brief encounters and struggles between ants of different species add up to the numbers and distribution of ants that are in a place at a certain time.

Interactions between particular species vary in outcome. Drawing on the metaphor of a pecking order or linear relation of power, familiar to us from human organizations, ecologists have used the notion of a dominance hierarchy to describe the relation among species that compete for food or other resources. But ants of a certain species don't always get the food first at the expense of the same other species. Instead, dominance hierarchies vary according to conditions, as Nathan Sanders showed in a guild of desert ants.[30] For example, Todd Palmer's work shows that among the four species that inhabit acacia trees in the African savanna, colony size determines which species gets the tree, presumably because larger colonies show up more often at resources and at sites of conflict, so that outcomes change as colonies grow.[31]

Ants interact with many other kinds of insects, which they use, protect, consume, or avoid. Ants' use of patterns of antennal contact with each other has carried over into their relations with other insects. For example, the caterpillars of many lycaenid butterflies are tended by ants. The caterpillars enter the ant nest, where the ants protect them. In turn, the larvae produce sugary liquids, or nectar, for the ants to eat. In many ant–lycaenid mutualisms, the ants stimulate the larvae to produce nectar by antennal drumming. Antennal contact between ants thus takes on a new function when practiced between ants and their mutualist partners, the caterpillars.

Antennal contact also mediates the relation of ants with their predators and with their prey. Mark Elgar and colleagues found that a jumping spider in the genus *Cosmophasis* uses both chemistry and behavior to prey on the larvae of the weaver ant, *Oecophylla smaragdina*.[32] The spider mimics the cuticular hydrocarbon profile of the colony, which allows the spider to enter the ant nest. When the spider finds an ant carrying a larva, it taps the ant, which drops the larva that the spider then takes out of the nest to eat. Other jumping spiders, of the genus *Myrmarachne*, have evolved to resemble ants and thus to avoid predation by ants. Here again, the mimicry extends to behavior. Fadia Ceccarelli discovered that the spiders wave two adjacent front legs in a way that resembles the antennal movements of ants.[33] It would be interesting to know how well the spider mimics are able to insert themselves into the ant network, and whether ants respond to the rate at which they meet the spiders in the same ways that they respond to similar rates of encounter with other ants.

Along with the relations every ant colony has with the other organisms that we can see, there is another world of ecological interactions among ants and organisms that we can't see, their pathogens and parasites. Dhruba Naug's work on honeybees shows how interaction networks determine the rate at which disease is transmitted.[34] A glimpse at the interactions of ants and pathogens comes from studies by Andrew Beattie and others of leaf-cutter ants.[35] Leaf-cutter ants rely for food on a fungus that they cultivate. The ants produce antibiotics to inhibit the growth of bacteria on their fungus. Ulrich Mueller and others have shown that the evolutionary history of many related species of leaf-cutter ants reflects the shifting associations between sets of species: the ants, the fungus that the ants eat, and the bacteria that also eat the fungus.[36] There are probably thousands of such webs of invisible interactions among ants and microorganisms that remain to be discovered. A whole new area of research will be to learn about the dynamics of these interactions.

While many species of ants eat fungi, some fungi consume ants. One fungal parasite uses the nesting and foraging behavior of the host ants to infect more ants. David Hughes and others study the fungus, *Ophiocordyceps unilateralis,* and the host ant *Camponotus leonardi,* in the tropical rain forest in Thailand. The fungus adds a macabre twist to the ants' behavior: an infected ant bites a leaf or stem and dies clamped on, while the fungus rapidly grows out of the body of the ant and begins to produce spores.[37] Ants get infected, not from each other, but by walking underneath or near these virulent spore-producing corpses. Because the ants return to a central nest, ants that were infected while foraging come back together and end up dying attached to leaves in the same neighborhood, creating a cluster of corpses. When a foraging trail passes underneath the cluster, many ants together are infected by the rain of spores. The fungus relies on the interaction networks that keep the ants together.

Invasive Species

The study of invasive species provides a unique opportunity to learn how ants interact with other species. An invasive species moves into a new community and changes it, because its behavior, how it uses resources to grow and spread, alters the existing use of resources by the native species. Examining how an invasive species changes a community sometimes can provide new insight on how that community works. For example, an analogous insight on humans would occur if, when the first woman professor joins a university science department, it suddenly becomes apparent that salacious jokes are frequent at department meetings. How a newly arrived population disrupts the network of interactions among colonies can help us guess what the network was like before the invaders arrived.

Ant invasions are not new, but like other ecological invasions, they have become much more frequent recently. In 1518 to 1520, an invasive ant, perhaps the fire ant *Solenopsis invicta*, overran the Carribean island of Hispaniola, destroying crops. It may be the same species of ant that became extremely abundant in Barbados in 1760 and exacerbated the effects of a disease of sugar cane. But in the past 30 years, transport of people and cargo, and the use of ballast water in large ships, have all accelerated enormously, and so has the inadvertent transport of propagules of many species.

There are a few species of ants that are invasive worldwide and do enormous ecological damage. The Argentine ant *Linepithema humile* traveled by boat with shipments of sugar from Argentina and has become established along the California coastline, the Mediterranean coastline, and parts of South Africa, Hawaii, and Australia. This severely diminishes populations of native ants, which in turn affects the distribution of native plants and arthropods and, indirectly, other animals. Argentine ants are considered an agricultural pest because they tend scale insects, which damage crops. Another worldwide invader is the tiny fire ant, *Wasmannia aurapuncta*, which has spread from South America through the South Pacific on timber and is also established in eastern Africa and Hawaii. *Pheidole megacephela* is invasive in many parts of the world, including the Caribbean, Africa, and Australia, and often competes successfully with other invasive ant species. Smaller invasions, successful only in one place, probably occur frequently. For example, the fungus-growing ant *Atta octospinosus* has colonized the Caribbean islands of Guadalupe. Sergey Mikheyev and others' genetic study shows that this species was introduced from Trinidad, Tobago, or northeast South America in the 1950s.[38] This species is not invasive elsewhere.

Invasion ecology began with the attempt to characterize what attributes of a species might predict its success as an invader. For example, some invasive ant species have small workers. Numbers rather than body size provide an advantage in ant warfare, and the little fire ant *Wasmannia auropunctata*, whose workers are already tiny in the native range, tends to have even smaller workers in the places it has invaded. Drawing on the argument that a colony can produce more workers when each worker is small, Terry McGlynn suggested that this species is evolving to have larger colonies of smaller ants in the areas where it is invasive.[39]

There is no single characteristic that makes invasive ants successful. Instead, their success arises from the local details of their network of ecological relations with other species. The outcome of invasions is the result of the accumulation of many encounters between the invader and local species. Some are direct, as when ants of a native species meet the invader at a source of food. Some are indirect—for example, when the invader uses up a resource that would have been available to the native ants if the invader were not there. The invasion comes to our notice when, eventually, these encounters tip the balance of so many ecological relationships between species that some species become locally extinct.

Ants of an invasive species can disrupt a community merely by showing up too often. Dennis Hansen and Christine Muller found that in Mauritius, the invasive ant *Technomyrmex albipes* harms the endangered endemic plant *Roussea simplex*, not just because the ant tends mealybugs that feed on the plant's fruits.[40] The plant depends on a lizard, the blue-tailed day-gecko *Phelsuma cepediana*, to pollinate its flowers and disperse its seeds. The ants hassle the lizards, so lizards leave plants when they meet ants of *T. albipes*. By keeping the lizard away, the ants make it harder for the plant to reproduce.

The success of the invasive Argentine ant in northern California is partly due to its interactions with us, and partly due to its interactions with native ant species. We are monitoring the spread of the Argentine ant in Jasper Ridge Biological Preserve, a reserve in northern California; this study is now in its seventeenth year.[41] Humans encourage Argentine ant populations to grow. We carry them to new places where their descendants form 'supercolonies,' genetically related groups of colonies. The supercolony probably makes no difference ecologically, but the structure of the individual colonies of Argentine ants helps them to make use of our buildings. Nicole Heller learned that in the winter, a colony aggregates in one or a few large nests, often at the base of shrubs in which the ants tend scale insects.[42] As the weather grows warmer in the spring, the ants begin to move out into distinct nests connected by trails. By the end of the summer, the colony is at its most dispersed, spanning about 200 meters, its many small nests connected by trails. Argentine ants come into buildings seeking water when it is hot and dry, and seeking warmth when it is cold and wet. Since the colonies are split into many nests with many queens, it is easy for a satellite nest with queens and workers to enter a building and to flourish while conditions are unfavorable outside. On the second or third cold, rainy day in the winter, I become fanatical about wiping off the kitchen counter, in a futile attempt to make it unappealing to exploring ants. When the weather changes, the ants leave their indoor nest and move back outside. All of our efforts to keep them out merely provide a temporary illusion of control; they leave when they are ready. Because we provide a refuge, Argentine ant populations are larger in developed areas.

The native ants are engaged in a genteel dance with each other that the Argentine ants exploit. In their native range, Argentine ant populations are small because of their competitive interactions with the local ants. In our study in northern California, we found that the outcome of an encounter between species depends on who shows up first.[43] When one species arrives at bait before another, the one that arrived second is likely to retreat. This is true of Argentine ants as well as native species. But Argentine ants are especially effective at adjusting their searching behavior to the number of available ants,[44] and so they may often find food first, causing the native species to retreat. Over time, this can starve out the native ants. However, some native species, especially the winter ant *Prenolepis imparis*, seem better able than others to persist when the Argentine ants show up. Chad Tillberg and others showed that Argentine ants rely more and more on scale insects for food as they become more established in an invaded area.[45] The winter ants also tend scale insects. We are currently investigating whether the winter ants can persist because they don't retreat when Argentine ants show up at their aphids.

The attempt to eradicate another famous invader, the red imported fire ant, *Solenopsis invicta*, revealed strong interactions among native ant species. From 1957 to 1982, millions of acres throughout the southeast United States were sprayed from the air with extremely potent pesticides, including dieldrin, heptachlor, and mirex, which were toxic to all ant species (as well as to small mammals, birds, and no doubt humans; these carcinogenic pesticides are now banned in the United States). The treated area was larger than the area in which fire ants had initially spread. Because fire ant populations recovered faster than those of native ant species, soon fire ants dominated ant communities throughout the treated region. Thus, the spread of fire ants was an unintended consequence of our impact on native species[46]: when we disrupted the interactions of fire ants with native species, populations of fire ants grew quickly. Eventually, native ants may recover and are able to suppress the fire ants' growth, as Lloyd Morrison showed in a reserve in Texas.[47] The fire ant has recently become established in California and has begun to spread in places where the invasive Argentine ant has reduced native ant populations. The outcome of this gruesome, yet ecologically fascinating, encounter between two invaders will depend on how fire ants and Argentine ants interact. Perhaps the two species will divide the available habitat, as it seems that the Argentine ant has done in Hawaii with the invasive *Pheidole megacephala*.[48]

From Ecology to Behavior

How neighboring colonies interact with each other determines how they get the resources they need to survive. Over time, these interactions will produce the spacing of colonies on the landscape—for example, if two colonies can't share a tree, eventually there will be only one in each tree. Behavioral interactions create the spacing of colonies, which in turn influences the availability of resources and thus colony growth, which both feed back to influence the frequency and outcome of behavioral interactions.

One example of the relation between interactions and spacing is the work of John Vandermeer, Ivette Perfecto, and others on the distribution of colonies of *Azteca instabilis* in trees planted to provide shade for a coffee plantation in Mexico.[49] Although the trees were planted in a regular array, the ant nests are clustered in small groups of trees. A computer simulation showed that this pattern would arise if mated ant queens fly off at random from mating aggregations and then some colonies establish satellite nests in nearby trees. Eventually, this process would lead to clusters of nests in huge groups of trees. But there is another factor at work: The size of clusters of trees with nests is limited by phorid flies. The flies lay their eggs in the heads of ants; as the egg grows into a larva, it kills the ant. Ants respond to the phorids by foraging less. Both by killing ants outright and by inhibiting foraging, which decreases the ants' food supply, the phorids keep clusters from growing larger. These processes in combination produce small clusters with relatively ant-free spaces between them. The configuration of the clusters in turn determines how much, and how often, colonies interact with each other, and how much they compete for food.

It takes only a short walk in a tropical forest to realize that every ant colony lives close to colonies of other species. There are ants of all shapes and sizes, on the ground, in the trees, on every leaf and twig and vine. I first visited the Amazon with Megan Frederickson, then a graduate student, who has studied Amazonian ants for many years. She was amused to see that I kept looking at the ground for ants, a habit formed over years of studying ants in the desert, while the ants rained down my neck from the vegetation above. I learned to look up.

I had been puzzled by the paucity of research on the behavior of tropical ants until I began to work in the tropics myself. We are beginning to learn about the diversity of ants in the tropics, as a result of collecting and identifying ants; this provides snapshots that show the outcome of relations among colonies. But we know little about everyday, ongoing relations among ant colonies. Often there are many species on the same plant. We do not yet see the patterns in these interactions. We do not know how often the ants of species A meet the ants of species B, whether particular colonies develop relations in which they recognize and adjust to each other, and how these interactions affect the ecology of each participant. On my first day in a tropical forest, I suddenly understood the reason for our ignorance. It's hard to stand under a tree full of stinging ants and calmly contemplate their comings and goings. The temptation is strong to count something quickly and get out of there, but we need studies that show how interactions play out over time.

Ants are crucial in all ecosystems on earth. Other animals eat ants, follow them to scavenge, depend on them for protection, or use their formic acid to clean off lice. Many plants rely on ants to disperse their seeds. Kathleen Treseder showed that ant breath is an important source of carbon for tropical epiphytic plants in which ants nest.[50] Ants transform soil everywhere, aerating it with their tunnels and enriching it with the nutrients they collect and absorb. Considering the importance of ants, we are remarkably ignorant about their ecology.

6

ANT EVOLUTION

Coevolution of Ants and Plants

The evolution of the ants is the story of how colony social organization has developed, expanded, and diversified in response to plants. The earliest ant fossils, from more than 130 million years ago,[1] resemble their wasp ancestors in many ways. About 90 million years ago, the ants began to diversify. It is clear that colony behavior in ants is intimately related to the evolution of diversity in plants.

Many early ant fossils belong to the Ponerine subfamily.[2] Existing ponerine species have small colonies that grow slowly, and many are predators that forage in the leaf litter; perhaps the ancestral ponerines did as well. Most of the plants at the time of these early ants were gymnosperms, including conifers, gingkos, and cycads. A huge asteroid wiped out much of life on earth, about 65 million years ago at the K–T boundary between the Cretaceous and Tertiary eras. By this time, the ants had already evolved into the subfamilies that survived the effects of the asteroid and persist today. The next burst of diversification of the ants occurred at the same time as the origin and radiation of the angiosperms, the flowering plants.

After the K–T boundary, the flowering plants diversified enormously, replacing many of the gymnosperms. In response, scale insects evolved to feed from plants, and the ant Formicine and Dolichoderine subfamilies evolved a wide array of relationships with scale insects. The groups of ants that tend and feed on scale insects now predominate in the tropical forest canopy.

Meanwhile, the Myrmicine subfamily developed large-colony species that forage on the ground, and these expanded into deserts, cooler forests, and arid grasslands. As ant diversity has exploded over the past 50 million years, ants have radiated into many small lineages, producing such a bushy evolutionary tree that among living ants, there is no clear trajectory from primitive to advanced.

While the evolution of new plant species provided many opportunities for new ant species to make a living, the ants also contributed to the diversity of plants. For example, many species of flowering plants deploy ants to disperse their seeds. A small packet of lipids, called an 'elaiosome,' is attached to each seed. The ants carry the seed home, eat the snack, and throw out the seed, which is still intact and able to grow. Szabi Lengyel and others showed that plant lineages with elaisomes on their seeds tend to have more species,[3] and suggest that by moving seeds, the ants reduce competition between plant parents and offspring, and that this has created opportunities for new plant species.

The many ant–plant mutualisms of the tropics provide an opportunity to learn how relations between ants and plants have shaped the behavior of the ants, because these mutualisms represent an extreme: For both the ant and plant mutualist partners, the ecological relationship is crucial. The plant partner in the mutualism produces a nesting place, and sometimes food, especially for ants, and the ant partner lives in the plant and defends it by attacking herbivorous insects that feed on the plant's leaves. There is an enormous diversity of ant–plant mutualisms in tropical forests, involving more than 40 ant and 100 plant genera; clearly, ant–plant mutualisms have arisen at different times in many different lineages.

Ant species that live in a mutualistic relationship with a plant manipulate the plant's biology. For example, some plant-ants prune vegetation growing around their host plants, which reduces competition for the host plant, allowing it to grow faster and thus to house more ants. Megan Frederickson learned that *Myrmelachista schumanni* ants climb onto any plant of another species, and then chew a hole in the stem of the alien plant and inject formic acid into its vascular system, causing the leaves to wither and the alien plant to die.[4] Alain Dejean and others found that *Allomerus decemarticulatus* ants build traps to capture insect prey on their host plant *Hirtella physophora*,[5] by cutting and binding plant hairs together, using a sooty-mold fungus they cultivate in the plant hairs. The ants then ambush and eat the insects that get caught in the sticky trap.

What is most remarkable about these examples is that the ants' behavior draws on very precise and intimate use of the plant's physiology. Most plant-ants have evolved refined and elaborate interactions with plants. The harvester ants I study in Arizona bring into the nest a motley collection of plant parts, of which only some turn out to be edible seeds. Anything faintly resembling a seed is good enough for them. In contrast, the *M. schumanni* are much more discriminating: they distinguish their host plants from all others, and they are able to find the particular sites on the stems of the other plants that will connect to the vascular system to spread the lethal formic acid.

What drives the evolution of the behavior of the plant-ants is that each partner has to adjust its behavior to the output of the other. Food and nest production by the plant are linked to defense by the ants. In many ant–plant mutualisms, there is a third partner, scale insects. Scales are common in ant–plants with domatia, swollen nodes of branches that contain a hollow space for ants to live in. The ants protect the scales that stick to the inside wall of the domatium and feed on the plant, excreting a sugary solution called 'honeydew' that the ants eat. Penny Gullan, a taxonomist who studies scale insects, pointed out to me that since each scale spends its whole life with its stylet stuck into the plant, sucking up phloem, the scales would drown in their own excrement if the ants did not drink it. Thus, ants use the scales for food, the scales need the ants to care for them, and the plant has to support the scales to keep the ants, who chase off insects that eat the plant.

The contributions of scales, ants, and plants are constantly being adjusted. The more the plant feeds the scales, the more ants will eat, and the faster the colony will grow. The more the colony grows, the better it can defend the plant, allowing more plant growth and presumably more resources available to feed the scales.

These happy exchanges among mutualist partners, what Bob May called "an orgy of mutual benefaction,"[6] vary in intensity as the ants and plants grow. Elizabeth Pringle, Rodolfo Dirzo, and I found, for example, that in the mutualism of *Cordia alliodora* and *Azteca* in Mexico, ants are more likely to enter young domatia.[7] We put caterpillars on leaves near young domatia and near old ones and found that the caterpillars near new domatia were attacked almost instantly. Overruling the more bloodthirsty inclinations of the undergraduates helping us, we decided to end each experiment after 5 minutes, by which time caterpillars near new domatia either succumbed to a horde of attacking ants or were forced to jump off the tree altogether to escape. But caterpillars near old domatia were often munching peacefully after 5 minutes, undisturbed by the ants. The young domatia are near new leaves, which in all plants tend to work harder than old ones at photosynthesis. Because the plant gets most of its food from the new leaves, the new leaves are more important than old ones for the plant. It seems that the ants are more likely to be in young domatia, and thus are available to defend the new leaves.

We are investigating the three-way set of interactions that channels the ants to the young domatia. One possibility is that the plant offers better phloem to the scales in the newer domatia, and this then attracts the ants, who end up being near the new leaves that the plant most needs to defend from herbivores. The ants and the plants are evolving together with the scales in the middle, because the growth of both ants and plants depends on the response of the scales.

Evolution of Colony Organization

One of the fascinating questions about ant evolution is how colony organization evolved from the behavior of ancestors who did not live in colonies. We have the same question about the evolution of any complex biological system. How did the coordination of all the cells in a multicellular organism arise, as separate, unicellular organisms fused and began to function together over millions of years of evolution? How did the web of interacting neurons in a vertebrate brain evolve from sets of more independent ganglia in the vertebrates' ancestors?

For behavioral traits like the organization of ant colonies, we have no fossil record and so cannot reconstruct the evolutionary history. A rich fossil record on the more recent vertebrates makes it possible to outline the trajectory that led polar bears to have thick fur to keep warm, or giraffes to have long necks to reach the tops of trees, but we know little about the steps on the way to the colony organization of ants.

Attempts to imagine the trajectory of the early evolution of ant colony organization have centered mostly on how much each colony member reproduces. One obvious feature of colony life, compared to the lives of the solitary wasps, the ants' closest ancestors, is that only one or a few females lay eggs, while the rest are sterile. For social insects, most of the thinking so far about the evolution of behavior has centered only on this trait: Who lays the eggs? Among the wasps and bees, there are examples of the two extremes—all females reproduce and then leave their eggs behind to fend for themselves, or only one female reproduces and her sterile daughters live with her to do the work of rearing the young—and every conceivable intermediate version. In most ant species, the queen or queens mate and lay eggs, with fertilized eggs that grow into females and unfertilized eggs that grow into haploid males. There are exceptions, species in which females arise from unfertilized eggs, or in which males arise from fertilized, diploid eggs and are sterile. Then there are species in which workers become queens—for example, Phil Ward showed that in some *Rhyti-doponera* colonies, workers mate and become queens, while in others, there is a single, morphologically distinct queen.[8]

The existence of sterile workers is a puzzle. If we consider sterility as a genetically inherited trait, it is hard to see how it could persist. Since any individual with this trait does not reproduce, how could the trait increase in frequency over generations? W. D. Hamilton came up with a brilliant answer to this question, drawing on the unusual genetic system of the Hymenoptera, the order of insects that includes ants, bees, and wasps.[9] In the Hymenoptera, females are diploid, with two sets of chromosomes, while the males are haploid, with only one set of chromosomes. As in other diploid organisms, like us, females are produced from eggs laid by females, fertilized by mating with a male. But the haploid males are produced from unfertilized eggs laid by females; usually males do not have fathers. Hamilton worked out that if a female queen mates with only one male, her daughters will be more closely related to their sisters than they would be to their own daughters. This is because all sisters will have in common the 50% of their genes that come from their father, since their haploid father contributes his genotype to all of them; his is half of the genotype of all his daughters. In addition, the sisters share on average half of the half, or 25%, of the genes of their diploid mother. This makes sisters, who are all daughters of the same mother and father, likely to share 50 plus 25, or 75% of their genes. However, diploid hymenopteran mothers and daughters, like other diploid parents and offspring (for example, humans), share only 50% of their genes, the half that the mother contributed to the daughter. Thus, Hamilton argued, if a female had genes that led to sterility, her sisters, with whom she has 75% of the genes in common, are more likely to have the same genes than her own daughters would be. Once worker sterility arose somehow, inclusive fitness or kin selection could promote the persistence of worker sterility, in which workers help their mothers to rear sisters.

Hamilton's idea about the consequences of haplodiploidy for the evolution of worker sterility, although brilliant, does not seem to account for how worker sterility persists in ants, because in most ant species, the queen mates with more than one male. When there is more than one patriline of daughter workers, many of a worker's sisters are in fact half-sisters, with the same mother but a different father. This leads workers to be *less* related to their sisters, sharing less than the 50% of their genes, than they would be with their own daughters— not more.

Presumably, worker sterility was present in ants from the beginning; at least, it is present in most ants we know today. Mary Jane West-Eberhard suggests that worker sterility arose originally in wasps, not because kin selection maintained it once it appeared out of the blue, but instead as a gradual evolutionary modification of other physiological processes in the ancestors of social insects. More generally, West-Eberhard's premise is that the variation upon which natural selection acts comes from developmental responses to the environment, rather than from genetic mutations that suddenly introduce new traits such as worker sterility.[10] Developmental responses determine how gene expression is triggered by the environment, and variation in developmental response leads to variation in phenotype. These genetic responses to environmental conditions are inherited, and over time, some may be favored by natural selection.

West-Eberhard, more recently James Hunt,[11] and others, have used these ideas to explain the evolution of sterile workers in wasps as the result of the evolution of environmental effects on egg-laying and the 'reproductive ground plan.' Wasps are a large and diverse group in which worker sterility has evolved many times. It seems that the regulation of reproduction, with flexibility that allows some wasps to become egg-layers when needed, has been ecologically important in many different times and places. Wasps show a variety of permutations: everyone lays eggs, or everyone could lay eggs but some females sometimes don't, or some lay eggs and some never do. It is clear that who lays the eggs is still labile in wasps—for example, if wasps of a species in which each female lives alone and lays eggs are all forced to live together, then some will lay eggs and some not. The reverse is also true—in species in which a group of females live together, and only one lays the eggs, the removal of that egg-laying female can trigger developmental changes so that a formerly sterile female starts to lay eggs. This means that interaction among workers determines who lays the eggs. Such flexibility is evidence against the notion that worker sterility arose suddenly from a genetic mutation, and evidence for the notion that instead, environmental conditions dictate whether a female lays eggs. Hunt shows that in wasps, diapause, seasonal changes in the rate of egg-laying, is linked to hormonal and other physiological processes, so that evolutionary modification of environmental effects on diapause could lead to worker sterility.

There remain many questions about the evolution of worker sterility in the wasps, and how it was modified in the transitions, over millions of years, that produced the ants from their wasp ancestors. Whatever the explanation for the origin and

widespread persistence of worker sterility in ants, it appears
that somehow, colonies with workers do better than groups
of reproducing females. As Darwin put it in his *Origin of Species*,[12] referring to a colony as a "family":

> I will here take only a single case, that of working or
> sterile ants. How the workers have been rendered sterile
> is a difficulty; but not much greater than that of any
> other striking modification of structure; for it can be
> shown that some insects and other articulate animals in
> a state of nature occasionally become sterile. . . . This
> difficulty, though appearing insuperable, is lessened,
> or, as I believe, disappears, when it is remembered that
> selection may be applied to the family, as well as to the
> individual, and may thus gain the desired end.

Whatever led to the evolution of worker sterility, there is
much more to the evolution of colony behavior. Who lays the
eggs does not determine the organization of the colony. The
fact that a worker is sterile does not dictate what she does.
Egg-laying is often portrayed as one side of a trade-off: either
a female lays eggs herself, or she works to raise her siblings,
her mother's offspring. In fact, many ant queens appear not
to do much besides laying eggs. But workers do a great many
things. Not laying eggs is not in itself a kind of behavior; it is
merely the absence of a certain behavior. A species in which
some individuals reproduce and others do not is called 'eusocial,' truly social. How egg-laying is distributed among individuals in a colony is sometimes called its 'social structure.'
This misleading language equates the allocation of egg-laying with social organization and directs attention away from

everything else that makes up the diverse and complex social organization of ant colonies. We wouldn't hope to explain how a nation's government works by identifying which government officials have children, or to describe the difference between a symphony orchestra and a rock band by contrasting how much time the players spent on maternity leave.

The social structure of an ant colony is much more than who lays the eggs, and the evolution of ant behavior is the evolution of coordinated colony organization. If a genie popped out of a bottle and offered to tell me anything about the origin of ant colonies, I would ask to learn about the trajectory that led ants to translate simple patterns of encounter into decisions about what to do next. This is what produced the organization of the colonies we see today.

Natural Selection in Action

It is not easy to find out what shapes the evolution of behavior. Over many generations, natural selection can produce changes in behavioral traits in the same way that it produces changes in size, shape, form, and physiology. But tracing the action of natural selection on behavior poses special problems. Natural selection requires inheritance, and the inheritance of behavior is especially puzzling. We know that developmental and environmental processes trigger the expression and the inhibition of genes and that somehow all of these processes lead to the traits we see in organisms. For behavior, the relation between the characteristics we see and the environment in which the behavior occurs is especially pronounced. Behavior is a response to environment; of course, so is anatomy, but it's not as obvious. We know little about the processes that regulate the inheritance of behavioral responses to changing conditions.

Along with inheritance, another necessary component of natural selection is a trend, over many generations, in relative reproductive success—the individuals with one version of a trait consistently reproduce more than others. It is always difficult to identify what causes some traits to improve reproductive success, because reproductive success is a result of ecological processes that affect many traits simultaneously. Again, because behavior is ephemeral, it is especially difficult to link behavior and reproductive success over an entire lifetime. Despite all of these challenges, we can try to learn how colony behavior is currently evolving, over a small number of generations, and ask how certain traits are related to reproductive success.

Variation, differences among individuals, is the source that feeds the action of natural selection. Unless individuals are different, there is no opportunity for natural selection to favor some relative to others. Thus, to investigate, in any existing population, how natural selection is shaping behavior, the starting point is to examine variation among individuals. Since in ants, it is the colonies that reproduce, natural selection on ant colony behavior acts on variation among colonies.

Interaction networks are the result of specific behavioral traits. How often ants meet depends on how each ant moves around, how quickly they move, and where ants of certain tasks go. For example, harvester ants go in and out of the nest on short trips, traveling to different destinations according to task: foragers travel on trails many meters from the nest, and nest maintenance workers put down their loads only a few centimeters from the entrance and then go back in. These behavioral traits have the consequence that ants interact mostly at the nest entrance between trips, and the longer the trip outside the nest, the longer the time elapsed until the next interaction back at the nest.

Natural selection could act on the function of interaction networks if within a population of colonies, some colonies use interactions in a slightly different way from others. For example, it seems that in some harvester ant colonies, the ants react a little more quickly than in others. How many interactions it takes to stimulate an ant to change its activity, and how long the effect of each interaction takes to decay, may vary among colonies.

I am currently studying whether natural selection is shaping the networks that regulate foraging behavior in harvester ants. As discussed in chapter 3, foragers respond very rapidly, within minutes, to changes in the rate at which successful foragers return, which is linked to food availability and how long the foragers had to search. This rapid response seems much faster than it needs to be. The invasive species that show up the second a crumb appears on your kitchen counter need to be opportunistic, because they rely on their ability to capture ephemeral resources. Harvester ants don't need to react within minutes to the availability of seeds scattered around the desert. The seeds are going to be in much the same place in 10 minutes and in 10 hours. But even though harvester ants don't need to be opportunistic, they do adjust to changes in foraging rate within minutes.

The rapid response of harvester ants to a change in food availability could be the result of selection to minimize the cost of foraging. Harvester ants obtain most of their water from metabolizing the fats in seeds. When they forage in the hot, dry desert air, they are using water to get more water. By slowing down foraging when food is hard to find, they cut down on the water they waste in fruitless searching. Decreasing the intensity of foraging rapidly when the stream of incoming foragers slows down could prevent the colony from wasting a molecule more of water than it has to. But that could just be an appealing story. The rapid response might not be due to any advantage in the rapid regulation of foraging. Instead, it could be merely a function of the short memory of a forager about its recent encounters with other ants, memories that fade rapidly after about 10 seconds.

Does it help a colony to be sensitive to food availability, slowing down its foraging at the first sign that it takes a long time to find food? Are colonies more successful if they make finer adjustments in the rate of foraging, relative to the rate at which food is coming in? Colonies that are more sensitive might get more food or waste less energy; on the other hand, it might be better for a colony to be more robust and keep going out to search no matter what.

We do know that colonies vary in foraging behavior.[13] All colonies respond to differences among days; maybe because of the weather, on some days most colonies forage, and on other days only the hardiest come out to forage. But some colonies forage every day, no matter what. Others seem much more finicky, foraging only on some days and sending out few foragers one day and many the next. This is not just due to differences in colony size, because there are colonies that have very large numbers of ants foraging but only on some days. These differences among colonies in foraging behavior persist from one year to the next. They are characteristic of colonies despite the turnover in ants. There does not seem to be any effect of location; we know that the distribution of food is ephemeral, and often there are both feisty and feeble colonies in the same area.

These differences among colonies in how much they forage could arise from differences in an individual's response to interactions, such as the threshold at which interaction activates a forager. Inactive foragers are stimulated to leave the nest when foragers come back to the nest with food. The higher the rate at which an inactive forager must meet returning ones before it leaves, the higher its threshold. Colonies might vary in such thresholds, and small variations could have large consequences for the behavior of the colony. If an ant requires 10 encounters within a certain time before it will leave the nest, it is less likely to leave the nest than if it requires only 2.

Another possibility that we are investigating is that colonies differ in the conditions under which the colony adjusts foraging intensity at all. Suppose that a forager returns to the nest and is likely to leave the nest on its next trip after some standard interval, unless it experiences a rate of interaction with returning foragers that is different from some baseline rate. For example, "If I meet a returning forager every 5 seconds, I am likely to go out again within 15 seconds. But if the returning foragers show up every 2 seconds, I am likely to go out again within 10 seconds." Colonies may vary in the magnitude of the increase or decrease in the rate at which foragers return needed to trigger a response. Then in one colony, the inactive foragers would respond to a small difference from

the baseline in their rate of encounter with returning foragers, while in another colony, it would require a larger change
to jolt the foragers from their standard waiting time. Overall, the more sensitive colony will adjust its foraging behavior
more often.

For natural selection to shape the responsiveness of colonies to food availability, variation among colonies would have
to be heritable, meaning that offspring colonies would resemble their parent colonies more than others in the population.
The colonies that are offspring of colonies with higher thresholds would be less sensitive than the offspring of colonies with
lower thresholds, willing to go out and forage at the drop of a
hat or just a few foragers carrying in seeds.

We know that there is variation among colonies in foraging intensity. The first condition for natural selection is met.
Almost nothing is known about the second condition, the
heritability of ant behavior. It's been suggested that the more
males a queen mates with, the more diverse will be the behavior of the workers, but apart from the work of Laura Snyder
on *Formica argentea*,[14] there are few data to support this. No
one has yet measured how much offspring colonies resemble
their parent colonies. When we can identify the offspring colonies of known parent colonies in harvester ants, we will see
whether the offspring of the more sensitive colonies resemble
their parents.

Last, if natural selection is shaping the regulation of foraging, the differences among colonies lead to a difference in
reproductive success. We are currently testing whether this
third condition for natural selection is met in the study population of harvester ants: whether variation in foraging behavior leads to variation in reproductive success. We measure

how sensitive colonies are and then see how many offspring they have.

Measuring reproductive success is tricky because of the way that harvester ants, and most other ants, reproduce. Winged reproductives, males and daughter queens, leave the nest to participate in a mating aggregation. The newly mated queens fly off at random to found new colonies. So offspring colonies do not end up in the same neighborhood as their parents, and there is no way to know which colonies produced the parents of a new colony. To find out how many offspring colonies each parent colony makes, we are using genetic methods, similar to DNA fingerprinting, to match up parents and offspring. We can then find out whether the more sensitive colonies have more offspring.

In the course of this work, we discovered by accident that harvester ants have a peculiar three-sex system.[15] We were developing markers for particular sequences of DNA that could be used to identify which colonies are offspring of which parents, an essential step for our studies of the evolution of foraging behavior. To our surprise, it turned out that one of the markers was always similar, or homozygous, for the maternal and paternal genes in queens, but always different, or heterozygous, in workers. This means that there must be two lineages within the population that mate with each other. When a queen mates with a male of the same lineage, the offspring are female reproductives. When she mates with a male of the other lineage, the offspring are sterile female workers that do not produce any mixed-lineage progeny. Thus, the queen must mate at least twice, once with a male of each lineage, to be able to found a viable colony with workers that can support the colony until it is large enough to produce daughter queens. Queens lay eggs of each type apparently in proportion to the number of males of each lineage that they mated with,[16] but the reproductive eggs do not grow into adults until the colony reaches some threshold size or food supply.[17] Thus, it seems that just as in other ant species and in honeybees, the feeding or care by the workers, who apparently can distinguish between reproductive and worker eggs, determines when and how many reproductives are produced.

Although the two genetic lineages in harvester ants are distinguished by a microsatellite marker, the marker itself designates a sequence that has no function. This marker must be linked with other genes, which we know nothing about, that are associated somehow with the reproductive status of the offspring.

We do not know how widespread such two-lineage systems are among ants generally. To me it seems unlikely that this would occur only in harvester ants. I suspect that when we examine the population genetics of more ant species, we will find many such systems. There may be a tendency in ants, as in some plants, to develop incompatibilities between the cytoplasm and nucleus of a cell, producing systems in which small mutations would eventually lead to the inability of each lineage to reproduce without the other.[18] In harvester ants, the existence of the two lineages influences the dynamics of evolution, because it affects the eligibility of each ant to mate with the others, and this in turn constrains the flow of genes from one generation to the next throughout the population. Although this system complicates the evolution of this population, it also makes it easier for us to identify which colonies are the offspring of others, because queens must be of the same lineage as their mothers.

Studying variation among colonies in reproductive success is a way to learn whether natural selection is tuning the interaction networks that set the reactivity of colonies to their environments. If the reproductive success of a harvester ant colony depends on its ability to get enough food and water, then colonies that adjust foraging effort to get more food without wasting water will produce more offspring. If water loss is not a problem, then colonies that are less sensitive, and that go out to forage no matter what, might collect more food and have more offspring.

Natural selection on interaction networks will shape how the colony reacts to changing conditions. Suppose that colonies of a plant-ant vary in how much they defend the tree and that such variation is heritable. Colonies might differ in the extent to which each ant reacts to alarm from other ants, so in some colonies the ants would respond more to the vibration caused when a caterpillar lands on a leaf. Such reactions could be heritable; if so, colonies that are offspring of the more reactive colonies would be more reactive. Selection could then favor plant-ants that defend their plants better, when decreased herbivory promotes plant growth enough that the benefit to the ants from the increased availability of nests outweighs the costs of running around chasing caterpillars off the tree. This would promote the evolution of a stronger response to interactions among ants.

Natural selection happens slowly, pushes back and forth and in different directions, and is extremely difficult to track as it is happening in nature. Before we can say anything general about the evolution of colony organization in ants, we will need detailed studies of many different species that show how interactions within and between colonies respond to changing ecological conditions.

7

MODELING ANT BEHAVIOR

Suppose that we want to build a general model of ant behavior as a way of summarizing what we know about how ant colonies are organized. Of course, there is no such thing as generic ant behavior. Different species of ants do the same things. They all make and repair a place where the colony lives. They all take in resources from the outside, pass them around a colony, and make more ants. But different species do these things in different ways.

The daily round of an ant colony is made up of large numbers of brief, simple interactions. The outcome is a miracle of fine-tuning. Within a day, interactions regulate how many ants go out, how much food they collect, how much is fed to larvae, and how much is stored. Over time, the regulation of all the colony's tasks determines how quickly the colony grows and how much impact it has on the other species it interacts with: the plants whose seeds it moves, the host plant it defends, the scale insects it tends, the fungus it grows, the other species whose brood it steals.

Any model of ant behavior has to have at least two levels. The first specifies how the workers interact within the colony to regulate the acquisition, processing, and distribution of resources. The second specifies how the internal processes of the colony connect to the colony's environment. The colony's behavior influences colony growth, and how colony growth affects all the other organisms that the colony uses and interacts with, which in turn will feed back on the colony's ability to grow. Although the first level is sometimes called 'behavior' and the second 'ecology,' they are clearly inseparable.

The first level, the regulation of colony activity, would detail how individual ants use local information to make immediate decisions about what to do next: which task to perform, and whether to perform it right now or rest. There must be conditions that stimulate ants to go out and collect resources. For example, in harvester ants, foraging is stimulated when ants return to the nest with food. The ants outside the nest encounter the food and bring it back. The more quickly they return with food, the more ants will be stimulated to go outside. This process links food availability, translated into ant experience as the time it takes to find food, to the intensity of foraging, by way of interactions between returning and inactive foragers. Another process regulates the distribution of food. For example, in fire ants, ants inside the nest engage in short bursts of feeding with any hungry larva they meet. When the worker's crop is empty, it does not feed any larvae. This process links food availability, translated into the numbers of workers with full bellies, to larval growth; when the foragers have found more food, and there are hungry larvae, the larvae get fed more.

Ant colony behavior is regulated by making patterns out of noisy and random ingredients. There is stochasticity, randomness and noise, both in how the ants react and in what they are reacting to. Whatever the stimulus, an ant's reactions are not deterministic. Watch any ant for a while, and you will see that an ant does not respond the same way every time to the same conditions. In every study of ant behavior, there are always many ants who do not act in the expected way. For example, a series of studies by Franks, Pratt, and others on nest moving in *Temnothorax*, discussed in chapter 3, explains how 60 to 75% of colonies choose their nests.[1] But about 25% of the colonies always do something different. Why?

A second important source of randomness in colony behavior is the variability of the rate of interaction that an ant experiences. Which ant meets another depends on the many small accidents and surprises that determine how long it takes for an ant to get from one place to another and end up close enough to another ant for them to interact. Anyone who has followed an ant will sympathize with Mark Twain's observation that an ant's indirect path carrying something back to the nest was "as bright a thing to do as it would be for me to carry a sack of flour from Heidelberg to Paris by way of Strasbourg steeple."[2]

Many decentralized systems use encounter rates, or networks of interactions, as the information that determines behavior. An intriguing and unresolved question is how such systems manage the stochasticity of interaction networks. How do decentralized systems combine variable input and imprecise response, yet manage to have the system respond correctly enough of the time that it can function?

The second level of a generic colony behavior model, the ecological level, would detail how the colony's interactions with the rest of the world determine its growth and reproduction, and how this changes the colony's environment, which in turn changes its interactions and feeds back on its growth and reproduction. For example, a colony's growth may be correlated with the amount of food the colony gets, so that when more food is available, it grows faster. At some point, the colony could grow so large that it uses up the available resources, or enters into competition with neighbors. This might lead to limits on its growth. Or, in a mutualistic system—for example, when ants defend host plants that provide nest sites—the growth of the colony could mean better defense of the host plant, promoting plant growth and thus producing more nest sites.

We are not yet able to describe fully, for any species, all of the processes that link a changing environment to ant behavior and then to colony response. Whenever we figure out one step for any particular species, it is tempting to jump to imagine that it is general for all ants. Eventually, when we know more about ants, it will be possible to consider the differences among species in the way that colonies are regulated. This diversity of regulation will reflect the enormous differences among ant species in ecology. What we have so far are only some of the pieces of many different puzzles; fragments of the picture for army ants, other pieces for harvester ants, others still for fire ants, and so on. In this early stage of research on ant behavior, we hope that when we find a piece that fits in one puzzle, it will go in much the same place in the others, but this doesn't have to be the case.

When we watch ants in a TV documentary, the ants behave with purpose and meaning. This is partly because animation has taught us to interpret the behavior of moving bodies on the screen as stories about individuals. It is partly because the genre plays up the amazing accomplishments of animals with stories constructed to provide brief, colorful, and impressive dramas. When you watch real ants, they don't look like they do on TV. You see a lot of bumbling around, a few ants going the wrong way, ants pulling an object in different directions. Yet ants are extraordinarily successful, and colonies do in fact accomplish astonishing feats, building, navigating, bringing resources in, and throwing them out. The achievements of colonies do not arise from the skill and determination of individual ants.

The colony isn't like clockwork, but it is ticking. Like a high school seething with text messages, it is a lively network of brief interactions. But in an important way, the ant colony is crucially different from the high school: no ant really knows what the message means. So the colony is more like a brain: neurons flash across the synapses, but no neuron knows what the brain is thinking, or that the signal means to notice the upper-left corner of the visual field or feel the little toe or decide what's for dinner. The pattern itself is the meaning that each interaction carries: it matters how often and from whom the text message arrives but not what it says.

What is most amazing about ant colonies is that such variable, noisy processes create a system that can accomplish so much. The system is turbulent in every way. The experience of each ant is variable, only loosely tuned to the state of the world, because the rate at which each ant meets others depends on so many small contingencies in how the ants happen to move around. The reaction of each ant to the pattern of encounters it experiences is imprecise, because ants don't count very carefully and because an ant's response to the same experience will be different from one time to the next. Yet in thousands of species, and so in thousands of different ways, ants bump into each other as they travel through an unpredictable world, and their inept reactions to these encounters produce millions of colonies that are making millions more.

NOTES

Chapter 1. The Ant Colony as a Complex System

1. Rodgers, D. M. 2008. *Debugging the Link between Social Theory and Social Insects.* Baton Rouge: Louisiana State University Press.

2. Butler, C. 1609. *The Feminine Monarchie, or the Historie of Bees.* Theatrum Orbis Terrarum; Da Capo Press.

3. Drouin, J.-M. 1992. L'image des sociétés d'insectes en France a l'époque de la révolution. *Revue de Synthèse.* 4:333–345.

4. Wheeler, W. M. 1910. *Ants.* New York: Columbia University Press. See Rodgers, D. M. *Debugging the Link between Social Theory and Social Insects* for the history of the term "superorganized."

5. White, T. H. 1977. *The Book of Merlyn.* London: Fontana/Collins.

6. Wilson, Edward O. 1968. The superorganism concept and beyond. In *Colloques internationaux du Centre National de la Recherche Scientifique,* M. Chauvin, M. Noirot, and P. P. Grassè, Eds., no. 173, pp. 27–38. Paris: CNRS. Quote from p. 7.

7. Holldobler, B., and E. O. Wilson. 1994. *Journey to the Ants.* Cambridge, MA: Belknap Press of Harvard University Press.

8. Oster, G. F., and E. O. Wilson. 1978. *Caste and Ecology in the Social Insects.* Princeton, NJ: Princeton University Press.

9. Wilson, E. O., N. I. Durlach, and L. M. Roth. 1958. Chemical releasers of necrophoric behavior in ants. *Psyche* 65(4):108–114.

10. Gordon, D. M. 1983. Dependence of necrophoric response to oleic acid on social context in the ant, *Pogonomyrmex badius*. *Journal of Chemical Ecology* 9:105–111.

A recent study of the Argentine ant, *Linepithema humile*, shows that chemicals on the ant's cuticle disappear within a few hours of its death, stimulating other ants to transport the corpse to the midden. Choe, D.-H., J. G. Millar, and M. K. Rust. 2009. Chemical signals associated with life inhibit necrophoresis in Argentine ants. *Proceedings of the National Academy of Sciences of the United States of America* 106:8251–8255.

11. Hofstadter, D. 1999. *Gödel Escher Bach: An Eternal Golden Braid*. New York: Basic Books.

Chapter 2. Colony Organization

1. If you find an ant and would like to know its species' name, try *www.antweb.org* and *myrmecos.net*, which show photos, mostly the beautiful photos of Alex Wild, of many ant genera. If you can't match your ant with a photo to your satisfaction, or if you want to be sure that you have the right name, you need to find an entomologist. The best place to start (in the United States) is the nearest land grant college; a list is available on the web at *www.higher-ed.org/resources/land_grant_colleges.htm*.

2. Davidson, D. W., S. R. Castro-Delgado, J. A. Arias, and J. Mann. 2006. Unveiling a ghost of Amazonian rain forests: *Camponotus mirabilis*, engineer of Guadua bamboo. *Biotropica* 38:653–660.

3. Fellers, J. H., and G. M. Fellers. 1976. Tool use in a social insect and its implications for competitive interactions. *Science* 192:70–72.

4. Yanoviak, S. P., R. Dudley, and M. Kaspari. 2005. Directed aerial descent in canopy ants. *Nature* 433:624–626.

5. For example, see Deneubourg, J., S. Aron, S. Goss, J. M. Pasteels, and G. Duerinck. 1986. Random behavior, amplification processes, and number of participants: how they contribute to the foraging properties of ants. *Physica D* 22: 176–186.

6. For example, see Sumpter, D.J.T., and S. C. Pratt. 2003. A modelling framework for understanding social insect foraging. *Behavioral Ecology and Sociobiology* 53:131–144.

7. Bulh, J., J. Gautrais, J.-L. Deneubourg, P. Kuntz, and G. Theraulaz. 2006. The growth and form of tunnelling networks in ants. *Journal of Theoretical Biology* 243:287–98; Jost, C., J. Verret, E. Casellas, J. Gautrais , M. Challet, J. Lluc, S. Blanco, M. J. Clifton, and G. Theraulaz. 2007. The interplay between a self-organized process and an environmental template: corpse clustering under the influence of air currents in ants. *Journal of the Royal Society Interface* 4:107–116.

8. Smith, A. 1776. *An Inquiry into the Nature and Causes of the Wealth of Nations.* London: W. Strahan and T. Cadell.

9. Wilson, E. O. 1984. The relation between caste ratios and division of labor in the ant genus *Pheidole* (Hymenoptera: Formicidae). *Behavioral Ecology and Sociobiology* 16:89–98.

Workers of different size may differ in flexibility. Christophe Lucas and Marla Sokolowski found that majors and minors of *Pheidole pallidula* respond differently to a situation that calls for defense. In majors but not minors, when defense is needed, the brain makes more PKG, a cGMP-dependent protein kinase, which reduces production of foraging protein PPFOR, a change that is associated with less foraging and more defensive behavior. In minors, the need for defense does not stimulate increased PKG production. Lucas, C., and M. B. Sokolowski. 2009. Molecular basis for changes in behavioral state in ant social behaviors. *Proceedings of the National Academy of Sciences of the United States of America* 106:6351–6356.

10. Wilson, E. O. 1980. Caste and division of labor in leaf-cutter ants (Hymenoptera: Formicidae: *Atta*). *Behavioral Ecology and Sociobiology* 7:157–165.

11. Feener, D. H., and K.A.G. Moss. 1990. Defense against parasites by hitchhikers in leaf-cutting ants—a quantitative assessment. *Behavioral Ecology and Sociobiology* 26:17–29.

12. Yackulic, C. B., and O. T. Lewis. 2007. Temporal variation in foraging activity and efficiency and the role of hitchhiking behaviour in the leaf-cutting ant, *Atta cephalotes. Entomologia Experimentalis et Applicata* 125:125–134.

13. Beshers, S. N., and J.F.A. Traniello. 1994. The adaptiveness of worker demography in the attine ant *Trachymyrmex septentrionalis. Ecology* 75:763–775.

14. Fjerdingstad, E. J., and R. H. Crozier. 2006. The evolution of worker caste diversity in social insects. *American Naturalist* 167:390–400.

15. Gordon, D. M. 1989. Caste and change in social insects. In *Oxford Surveys in Evolutionary Biology*, P. D. Harvey and L. Partridge, Eds., pp. 56–72. Oxford: Oxford University Press.

16. Gordon, D. M. 2000. *Ants at Work: How an Insect Society Is Organized*. New York: W. W. Norton and Co.

17. Gordon, D. M. 1989. Dynamics of task switching in harvester ants. *Animal Behaviour* 38:194–204.

18. Sturgis, S., and M. J. Greene. Hydrocarbon labels on harvester ant nest mounds. In preparation.

19. Once an ant ecloses from the pupa, it is an adult, but its brain may continue to mature, allowing its behavior to develop throughout its life. Marc Seid and Rudiger Wehner found that in the ant *Cataglyphis albicans*, which is known for its talent for navigating desert landscapes, axonal pruning in the brain continues even after the adult ant is two months old. Seid, M. A., and R. Wehner. 2009. Delayed axonal pruning in the ant brain: a study of developmental trajectories. *Developmental Neurobiology* 69:350–364.

20. Franks, N. R., and C. Tofts. 1994. Foraging for work: how tasks allocate workers. *Animal Behaviour* 48:470–472.

21. Holldobler, B., and E. O. Wilson. 1990. *The Ants*. Cambridge, MA: Belknap Press of Harvard University Press.

22. Reviewed in Gordon D. M., J. Chu, A. Lillie, M. Tissot, and N. Pinter. 2005. Variation in the transition from inside to outside work in the red harvester ant *Pogonomyrmex barbatus*. *Insectes Sociaux* 52:212–217.

23. Seid, M. A., and J.F.A. Traniello. 2006. Age-related repertoire expansion and division of labor in *Pheidole dentata* (Hymenoptera: Formicidae): a new perspective on temporal polyethism and behavioral plasticity in ants. *Behavioral Ecology and Sociobiology* 60:631–644.

24. Muscedere, M. L., T. A. Willey, and J. F. A. Traniello. 2009. Age and task efficiency in the ant *Pheidole dentata*: young minor workers are not specialist nurses. *Animal Behaviour* 77:911–918.

25. Tripet, F., and P. Nonacs. 2004. Foraging for work and age-based polyethism: the roles of age and previous experience on task choice in ants. *Ethology* 110:863–877.

26. McDonald, P., and H. Topoff. 1985. Social regulation of behavioral development in the ant, *Novomessor albisetosus* (Mayr). *Journal of Comparative Psychology* 99:3–14.

27. Gordon, D. M., J. Chu, A. Lillie, M. Tissot, and N. Pinter. 2005. *Insectes Sociaux* 52:212–217.

28. Aron, S., J. M. Pasteels, and J. L. Deneubourg. 1989. Trail-laying behavior during exploratory recruitment in the Argentine ant, *Iridomyrmex humilis* (Mayr). *Biology of Behaviour* 14:207–217.

29. Greene, M. J., and D. M. Gordon. 2007. How patrollers set foraging direction in harvester ants. *American Naturalist* 170:943–948.

30. For example, Gordon, D. M. 1986. The dynamics of the daily round of the harvester ant colony. *Animal Behaviour* 34:1402–1419.

31. Gordon, D. M. 1987. Group-level dynamics in harvester ants: young colonies and the role of patrolling. *Animal Behaviour* 35:833–843.

Chapter 3. Interaction Networks

1. Ratnieks, F.L.W., and C. Anderson. 1999. Task partitioning in insect societies. *Insectes Sociaux* 46:95–108.

2. Wilson, E. O. 1985. Between-caste aversion as a basis for division of labor in the ant *Pheidole pubiventris* (Hymenoptera: Formicidae). *Behavioral Ecology and Sociobiology* 17:35–37.

3. Liang, D., and J. Silverman. 2000. "You are what you eat": diet modifies cuticular hydrocarbons and nestmate recognition in the Argentine ant, *Linepithema humile*. *Naturwissenschaften* 87:412–416.

4. Suarez, A. V., D. A. Holway, and N. D. Tsutsui. 2008. Genetics and behavior of a colonizing species: the invasive Argentine ant. *American Naturalist* 172:S72–S84.

5. Gordon, D. M. 2002. The regulation of foraging activity in red harvester ant colonies. *American Naturalist* 159:509–518.

6. Greene, M. J., and D. M. Gordon. 2007. How patrollers set foraging direction in harvester ants. *American Naturalist* 170:943–948.

7. Bonavita-Courgourdan, A., J. Clement, and A. Poveda. 1990. Les hydrocarbures cuticulaires et les processus de reconnaissance chez les fourmis: le code d'information complexe de *Camponotus vagus*. *Insectes Sociaux* 6:273–280.

8. Wagner, D., M.J.F. Brown, P. Broun, W. Cuevas, L. E. Moses, D. L. Chao, and D. M. Gordon. 1998. Task-related differences in the cuticular hydrocarbon composition of harvester ants, *Pogonomyrmex barbatus*. *Journal of Chemical Ecology* 24: 2021–2037. Wagner, D., M. Tissot, and D. M. Gordon. 2001. Task-related environment alters the cuticular hydrocarbon composition of harvester ants. *Journal of Chemical Ecology* 27:1805–1819.

9. Greene, M. J., and D. M. Gordon. 2003. Cuticular hydrocarbons inform task decisions. *Nature* 423:32.

10. Greene, M. J., and D. M. Gordon. 2007. Interaction rate informs harvester ant task decisions. *Behavioral Ecology* 18:451–455.

11. Schafer, R. J., S. Holmes, and D. M. Gordon. 2006. Forager activation and food availability in harvester ants. *Animal Behaviour* 71:815–822.

12. Greene, M. J., and D. M. Gordon. 2009. Combined social and seed chemical cues inform harvester ant foraging decisions. In review.

13. Beverly, B., H. McLendon, S. Nacu, S. Holmes, and D. M. Gordon. 2009. *Behavioral Ecology* 20:633–638.

14. Gordon, D. M., S. Holmes, and S. Nacu. 2008. The short-term regulation of foraging in harvester ants. *Behavioral Ecology* 19:217–222.

15. Pratt, S. C. 2005. Behavioral mechanisms of collective nest-site choice by the ant *Temnothorax curvispinosus*. *Insectes Sociaux* 52:383–392.

16. Wilson, E. O. 1962. Chemical communication among workers of the fire ant *Solenopsis saevissima* (Fr. Smith). I. The organization of mass foraging. *Animal Behaviour* 10:134–147.

17. Gordon, D. M., and N. Mehdiabadi. 1999. Encounter rate and task allocation in harvester ants. *Behavioral Ecology and Sociobiology* 45:370–77.

18. Page, R. E., J. Erber, and M. K. Fondrk. 1998. The effect of genotype on response thresholds to sucrose and foraging behavior of honey bees (*Apis mellifera* L.). *Journal of Comparative Physiology A* 182:489–500.

19. Jeanne, R. L. 1986. The organization of work in *Polybia occidentalis*: costs and benefits of specialization in a social wasp. *Behavioral Ecology and Sociobiology* 19:333–342.

20. Seeley, T. D. 1989. Social foraging in honey bees: how nectar foragers assess their colony's nutritional status. *Behavioral Ecology and Sociobiology* 24:181–199.

21. Dupuy, F., J. C. Sandoz, M. Giurfa, and R. Josens. 2006. Individual olfactory learning in *Camponotus* ants. *Animal Behaviour* 72:1081–1091.

22. Reznikova, Z., and B. Ryabko. 1994. Experimental study of ants' communication system, with the application of the Information Theory approach. *Memorabilia Zoologica* 48:219–236.

23. Beverly, B., H. McLendon, S. Nacu, S. Holmes, and D. M. Gordon. 2009. How site fidelity leads to individual differences in the foraging activity of harvester ants. *Behavioral Ecology* 20:633–638.

24. Gordon, D. M. 1991. Behavioral flexibility and the foraging ecology of seed-eating ants. *American Naturalist* 138:379–411; and Greene M. J., and D. M. Gordon. 2007. *American Naturalist* 170:943–948.

25. Fletcher, D.J.C., and M. S. Blum. 1983. Regulation of queen number by workers in colonies of social insects. *Science* 219:312–314.

26. Errard, C. 1994. Long-term-memory involved in nestmate recognition in ants. *Animal Behaviour* 48:263–271.

27. Rosengren, R. 1977. Foraging strategy of wood ants (Formica rufa group). I. Age polyethism and topographic tradition. *Acta Zoologica Fennica* 149:1–30.

28. Forel, A. 1904. *Ants and Some Other Insects.* Chicago: The Open Court Publishing Company.

29. Topoff, H., D. Bodoni, P. Sherman, and L. Goodloe. 1987. The role of scouting in slave raids by *Polyergus breviceps* (Hymenoptera: Formicidae). *Psyche* 94:261–270.

30. Jaisson, P., D. Fresneau, and J.-P. Lachaud. 1988. Individual traits of social behavior in ants. In *Interindividual Behavioral Variability in Social Insects*, R. L. Jeanne, Ed., pp. 1–52. Boulder, CO: Westview Press.

31. Beverly, B., H. McLendon, S. Nacu, S. Holmes, and D. M. Gordon. 2009. *Behavioral Ecology* 20:633–638.

32. Detrain, C., and J. L. Deneubourg. 1997. Scavenging by *Pheidole pallidula*: a key for understanding decision-making systems in ants. *Animal Behaviour* 53:537–547.

33. Gordon, D. M., R. Rosengren, and L. Sundstrom. 1992. The allocation of foragers in red wood ants. *Ecological Entomology* 17:114–120.

34. Fourcassie, V., and J.F.A. Traniello. 1994. Food searching behavior in the ant *Formica schaufussi* (Hymenoptera, Formicidae): response of naive foragers to protein and carbohydrate food. *Animal Behaviour* 48:69–79.

35. Cole, B. 1990. Short-term activity cycles in ants: generation of periodicity by worker interaction. *American Naturalist* 137:244–259; and Gordon, D. M. 1991. Comment on Cole, B. 1990. *American Naturalist* 137:260–261.

36. Adler, F. R., and D. M. Gordon. 1992. Information collection and spread by networks of patrolling ants. *American Naturalist* 140:373–400.

37. Gordon, D. M. 1995. The expandable network of ant exploration. *Animal Behaviour* 50:995–1007.

38. Gordon, D. M., R. E. Paul, and K. Thorpe. 1993. What is the function of encounter patterns in ant colonies? *Animal Behaviour* 45:1083–1100.

39. Pacala, S. W., D. M. Gordon, and H.C.J. Godfray. 1996. Effects of social group size on information transfer and task allocation. *Evolutionary Ecology* 10:127–165.

40. Dussoutour, A., S. Beshers, J.-L. Deneubourg, and V. Fourcassie. 2009. Priority rules govern the organization of traffic on foraging trails under crowding conditions in the leaf-cutting ant *Atta colombica. Journal of Experimental Biology* 212:489–505.

41. Gordon, D. M., R. E. Paul, and K. Thorpe. 1993. *Animal Behaviour* 45:1083–1100.

42. Gotzek, D., and K. G. Ross. 2007. Genetic regulation of colony social organization in fire ants: an integrative overview. *Quarterly Review of Biology* 82:201–226.

Chapter 4. Colony Size

1. Gordon, D. M., and A. W. Kulig. 1996. Founding, foraging and fighting: colony size and the spatial distribution of harvester ant nests. *Ecology* 77:2393–2409.

2. Tschinkel, W. R. 2006. *The Fire Ants*. Cambridge, MA: The Belknap Press of Harvard University Press.

3. Gordon, D. M. 1991. Behavioral flexibility and the foraging ecology of seed-eating ants. *American Naturalist* 138:379–411.

4. Gordon, D. M., and A. W. Kulig. 1998. The effect of neighboring colonies on mortality in harvester ants. *Journal of Animal Ecology* 67:141–148.

5. Gordon, D. M. 1992. How colony growth affects forager intrusion between neighboring harvester ant colonies. *Behavioral Ecology and Sociobiology* 31:417–427.

6. Cassill, D. L., and W. R. Tschinkel. 1995. Allocation of liquid food to larvae via trophallaxis in colonies of the fire ant, *Solenopsis invicta*. *Animal Behaviour* 50:801–813.

7. Schneirla, T. C. 1938. A theory of army-ant behavior based upon the analysis of activities in a representative species. *Journal of Comparative Psychology* 25:51–90.

8. Frederickson, M. E., and D. M. Gordon. 2009. The intertwined population biology of two Amazonian myrmecophytes and their symbiotic ants. *Ecology* 90:1595–1607.

9. Adler, F. R., and D. M. Gordon. 2003. Optimization, conflict, and nonoverlapping foraging ranges in ants. *American Naturalist* 162:529–543.

10. Mailleux, A. C., J. L. Deneubourg, and C. Detrain. 2003. How does colony growth influence communication in ants? *Insectes Sociaux* 50:24–31.

11. Thomas, M. L., and M. A. Elgar. 2003. Colony size affects division of labour in the ponerine ant *Rhytidoponera metallica*. *Naturwissenschaften* 90:88–92.

12. Gordon, D. M. 1987. Group-level dynamics in harvester ants: young colonies and the role of patrolling. *Animal Behaviour* 35:833–843.

13. Gordon, D. M. 1996. The organization of work in social insect colonies. *Nature* 380:121–124.

14. Pacala, S. W., Gordon, D. M., and H.C.J. Godfray. 1996. Effects of social group size on information transfer and task allocation. *Evolutionary Ecology* 10:127–165.

15. If you want to try this, you need to know about Stefan Cover's discovery that Keebler Pecan Sandies cookies make the best all-purpose ant bait because they have something for any ant species—protein, fat, and sugar.

16. Daly-Schveitzer, S., G, Beugnon, and J.-P. Lachaud. 2007. Prey weight and overwhelming difficulty impact the choice of retrieval strategy in the neotropical ant *Gnamptogenys sulcata*. *Insectes Sociaux* 54:319–328.

17. Fielde, A. M. 1904. Power of recognition among ants. *Biological Bulletin* 7:227–250.

18. Newey, P. S., S.K.A. Robson, and R. H. Crozier. 2008. Temporal variation in recognition cues: implications for the social life of weaver ants *Oecophylla smaragdina*. *Animal Behaviour* 77:481–488.

19. Anderson, C., and F. L. Ratnieks. 1999. Task partitioning in insect societies. I. Effect of colony size on queueing delay and colony ergonomic efficiency. *American Naturalist* 154:521–535.

20. For example, see DeHeer, C. J., V. L. Backus, and J. M. Herbers. 2001. Sociogenetic responses to ecological variation in the ant *Myrmica punctiventris* are context dependent. *Behavioral Ecology and Sociobiology* 49:375–386.

21. Heller, N. E., and D. M. Gordon. 2006. Seasonal spatial dynamics and causes of nest movement in colonies of the invasive Argentine ant (*Linepithema humile*). *Ecological Entomology* 31:499–510.

Chapter 5. Relations with Neighbors

1. Gordon, D. M. 1993. The spatial scale of seed collection by harvester ants. *Oecologia* 95:479–487.

2. Gordon, D. M. 2000. *Ants at Work: How an Insect Society Is Organized*. New York: W. W. Norton.

3. Gordon, D. M., and A. W. Kulig. 1996. Founding, foraging and fighting: colony size and the spatial distribution of harvester ant nests. *Ecology* 77:2393–2409.

4. Sanders, N. J., and D. M. Gordon. 2003. Resource-dependent interactions and the organization of desert ant communities. *Ecology* 84:1024–1031.

5. Gordon, D. M, and A. W. Kulig. 1996. How colony growth affects forager intrusion between neighboring harvester ant colonies. *Ecology* 77:2393–2409.

6. Gordon, D. M. 1992. *Behavioral Ecology and Sociobiology* 31:417–427.

7. Gordon, D. M. 1991. Behavioral flexibility and the foraging ecology of seed-eating ants. *American Naturalist* 138:379–411; and Greene, M .J., and D. M. Gordon. 2007. How patrollers set foraging direction in harvester ants. *American Naturalist* 170:943–948.

8. Gordon, D. M. 1992. *Behavioral Ecology and Sociobiology* 31:417–427.

9. Adler, F. R., and D. M. Gordon. 2003. Optimization, conflict, and nonoverlapping foraging ranges in ants. *American Naturalist* 162:529–543.

10. Roulston, T. H., G. Buczkowski, and J. Silverman. 2003. Nestmate discrimination in ants: effect of bioassay on aggressive behavior. *Insectes Sociaux* 50:151–159.

11. Carlin, N. F. 1989. Discrimination between and within colonies of social insects: two null hypotheses. *Netherlands Journal of Zoology* 39:86–100.

12. Greenberg, L. 1988. Kin recognition in the sweat bee, *Lasioglossum zephyrum*. *Behavior Genetics* 18:425–438.

13. Vander Meer, R. K., D. Saliwanchik, and B. Lavine. 1989. Temporal changes in colony cuticular hydrocarbon patterns of *Solenopsis invicta*: implications for nestmate recognition. *Journal of Chemical Ecology* 15:2115–2125.

14. Gordon, D. M. 1989. Ants distinguish neighbors from strangers. *Oecologia* 81:198–200.

15. Brown, M.J.F., and D. M. Gordon. 1997. Individual specialisation and encounters between harvester ant colonies. *Behaviour* 134:849–866.

16. Ozaki, M., A. Wada-Katsumata, K. Fujikawa, M. Iwasaki, F. Yokohari. Y. Satoji, T. Nisimura, and R. Yamaoka. 2005. Ant nestmate and non-nestmate discrimination by a chemosensory sensillum. *Science* 309:311–314.

17. Langen, T. A., F. Tripet, and P. Nonacs. 2000. The red and the black: habituation and the dear-enemy phenomenon in two desert *Pheidole* ants. *Behavioral Ecology and Sociobiology* 48:285–292.

18. Van Wilgenburg, E., D. Ryan, P. Morrison, P. J. Marriott, and M. A. Elgar. 2006. Nest- and colony-mate recognition in polydomous colonies of meat ants (*Iridomyrmex purpureus*). *Naturwissenschaften* 93:309–314.

19. Van Wilgenburg, E., S. Dang, A. L. Forti, T. J. Koumoundouros, A. Ly, and M. A. Elgar. 2007. An absence of aggression between non-nestmates in the bull ant *Myrmecia nigriceps*. *Naturwissenschaften* 94:787–790.

20. Cherix, D. 1980. Note préliminaire sur la structure, la phenologie et le régime alimentaire d'une super-colonie de *Formica lugubris*. *Insectes Sociaux* 27:226–236.

21. Pamilo, P., P. Getsch, and P. Seppa. 1997. Molecular population genetics of social insects. *Annual Review of Ecology and Systematics* 28:1–25.

22. Adams, E. S. 1994. Territory defense by the ant *Azteca trigona*: maintenance of an arboreal ant mosaic. *Oecologia* 97:202–208.

23. Gordon, D. M., R. E. Paul, and K. Thorpe. 1993. What is the function of encounter patterns in ant colonies? *Animal Behaviour* 45:1083–1100.

24. Franks, N. R., and L. W. Partridge. 1993. Lanchester battles and the evolution of combat in ants. *Animal Behaviour* 45:197–199.

25. Sagata, K., and P. J. Lester. 2009. Behavioural plasticity associated with propagule size, resources, and the invasion success of the Argentine ant *Linepithema humile*. *Journal of Applied Ecology* 46:19–27.

26. Gordon, D. M. 1988. Nest-plugging: interference competition in desert ants (*Novomessor cockerelli* and *Pogonomyrmex barbatus*). *Oecologia* 75:114–118.

27. Sanders, N. J., and D. M. Gordon. 2004. The interactive effects of climate and interspecific neighbours on mortality of red harvester ants. *Ecological Entomology* 29:632–637.

28. Adams, E. S. 1990. Interaction between the ants *Zacryptocerus maculatus* and *Azteca trigona*: interspecific parasitization of information. *Biotropica* 22:200–206.

29. Swain, R. B. 1980. Trophic competition among parabiotic ants. *Insectes Sociaux* 27:377–390.

30. Sanders, N. J., and D. M. Gordon. 2003. Resource-dependent interactions and the organization of desert ant communities. *Ecology* 84:1024–1031.

31. Palmer, T. M. 2004. Wars of attrition: colony size determines competitive outcomes in a guild of African acacia ants. *Animal Behaviour* 68:993–1004.

32. Allan, R. A., R. J. Capon, W. V. Brown, and M. A. Elgar. 2002. Mimicry of host cuticular hydrocarbons by salticid spider *Cosmophasis bitaeniata* that preys on larvae of tree ants *Oecophylla smaragdina*. *Journal of Chemical Ecology* 28:835–848.

33. Ceccarelli, F. S. 2008. Behavioral mimicry in *Myrmarachne* species (Araneae, Salticidae) from North Queensland, Australia. *Journal of Arachnology* 36:344–351.

34. Naug, D., and B. Smith. 2007. Experimentally induced change in infectious period affects transmission dynamics in a social group. *Proceedings Royal Society B* 274:61–65.

35. Stow, A., and A. Beattie. 2008. Chemical and genetic defenses against disease in insect societies. *Brain, Behavior, and Immunity* 22:1009–1013.

36. Currie, C. R., B. Wong, A. E. Stuart, T. R. Schultz, S. A. Rehner, U. G. Mueller, G. H. Sung, J. W. Spatafora, and N. A. Straus. 2003. Ancient tripartite coevolution in the attine ant–microbe symbiosis. *Science* 299:386–388.

37. Pontoppidan, M. B., W. Himaman, N. L. Hywel-Jones, J. J. Boomsma, and D. P. Hughes. 2009. Graveyards on the move: the spatio-temporal distribution of dead *Ophiocordyceps*-infected ants. PloS ONE 4(3): e4835.

38. Mikheyev, A. S., U. G. Mueller, and P. Abbot. 2006. Cryptic sex and many-to-one coevolution in the fungus-growing ant symbiosis. *Proceedings of the National Academy of Sciences of the United States of America* 103:10702–10706.

39. McGlynn, T. P. 1999. Non-native ants are smaller than related native ants. *American Naturalist* 154:690–699.

40. Hansen, D. M., and C. B. Muller. 2009. Invasive ants disrupt gecko pollination and seed dispersal of the endangered plant *Roussea simplex* in Mauritius. *Biotropica* 41:202–208.

41. Sanders, N. J., K.E. Barton, and D. M. Gordon. 2001. Dynamics of the distribution and impact of the invasive Argentine ant, *Linepithema humile*, in northern California. *Oecologia* 127:123–130; and Heller, N. E., N. J. Sanders, J. W. Shors, and D. M. Gordon. 2008. Rainfall facilitates the spread, and time alters the impact of the invasive Argentine ant. *Oecologia* 155:385–395.

42. Heller, N. E., and D. M. Gordon. 2006. Seasonal spatial dynamics and causes of nest movement in colonies of the invasive Argentine ant (*Linepithema humile*). *Ecological Entomology* 31:499–510.

43. Human, K. G., and D. M. Gordon. 1996. Exploitation and interference competition between the invasive Argentine ant, *Linepithema humile*, and native ant species. *Oecologia* 105:405–412.

44. Gordon, D. M. 1995. The expandable network of ant exploration. *Animal Behaviour* 50:995–1007.

45. Tillberg, C. V., D. A. Holway, E. G. LeBrun, and A. V. Suarez. 2007. Trophic ecology of invasive Argentine ants in their native and introduced ranges. *Proceedings of the National Academy of Sciences of the United States of America* 52:20856–20861.

46. Buhs, J. B. 2004. *The Fire Ant Wars: Nature, Science, and Public Policy in Twentieth-Century America*. Chicago: University of Chicago Press.

47. Morrison, L. W. 2002. Long-term impacts of an arthropod-community invasion by the imported fire ant, *Solenopsis invicta*. *Ecology* 83:2337–2345.

48. Krushelnycky, P. D., L. L. Loope, and N. J. Reimer. 2005. The ecology, policy, and management of ants in Hawaii. *Proceedings of the Hawaiian Entomological Society* 37:1–25.

49. Perfecto, I., and J. Vandermeer. 2008. Spatial pattern and ecological process in the coffee agroforestry system. *Ecology* 89:915–920.

50. Treseder, K. K., D. W. Davidson, and J. R. Ehleringer. 1995. Absorption of ant-provided carbon-dioxide and nitrogen by a tropical epiphyte. *Nature* 375:137–139.

Chapter 6. Ant Evolution

1. For example, see Brady, S. G., T. R. Schultz, B. L. Fisher, and P. S. Ward. 2006. Evaluating alternative hypotheses for the early evolution and diversification of ants. *Proceedings of the National Academy of Sciences of the United States of America* 103:18172–18177.

2. Wilson, E. O., and B. Holldobler. 2005. The rise of the ants: a phylogenetic and ecological explanation. *Proceedings of the National Academy of Sciences of the United States of America* 102:7411–7414.

3. Lengyel, S., A. D. Gove, A. M. Latimer, J. D. Majer, R. R. Dunn. 2009. Ants sow the seeds of global diversification in flowering plants. *PloS ONE* 4(5): e5480.

4. Frederickson, M. E., M. J. Greene, and D. M. Gordon. 2005. "Devil's gardens" bedevilled by ants. *Nature* 437:495–496.

5. Dejean, A., P. J. Solano, J. Ayroles, B. Corbara, and J. Orivel. 2005. Arboreal ants build traps to capture prey. *Nature* 434:973.

6. May, R. 1982. Mutualistic interactions among species. *Nature* 296:803–804.

7. Pringle, E., R. Dirzo, and D. M. Gordon. The role of symbiotic hemipterans in ant defense of a neotropical myrmecophyte. In preparation.

8. Ward, P. S. 1986, Functional queens in the Australian greenhead ant *rhytidoponera-metallica* hymenoptera formicidae. *Psyche* 93:1–12.

9. Hamilton, W. D. 1964. The genetical evolution of social behavior. II. *Journal of Theoretical Biology* 7:17–52.

10. West-Eberhard, M. J. 2003. *Developmental Plasticity and Evolution*. Oxford: Oxford University Press.

11. Hunt, J. H. 2007. *The Evolution of Social Wasps*. Oxford: Oxford University Press.

12. Darwin, C. D. 1964 *Origin of Species*, pp. 237–238. Cambridge, MA: Harvard University Press.

13. Gordon, D. M. 1991. Behavioral flexibility and the foraging ecology of seed-eating ants. *American Naturalist* 138:379–411.

14. Snyder, L. E. 1993. Nonrandom behavioral interactions among genetic subgroups in a polygynous ant. *Animal Behaviour* 46:431–439.

15. Volny, V. P., and D. M. Gordon. 2002. Genetic basis for queen–worker dimorphism in a social insect. *Proceedings of the National Academy of Sciences of the United States of America* 99:6108–6111.

16. Volny, V. P., M. J. Greene, and D. M. Gordon. 2006. Brood production and lineage discrimination in a harvester ant population with genetic caste determination. *Ecology* 87:2194–2200.

17. Gordon, D. M. 1995. The development of an ant colony's foraging range. *Animal Behaviour* 49:649–659.

18. Linksvayer, T., M. J. Wade, and D. M. Gordon. 2006. Genetic caste determination in harvester ants: possible origin and maintenance by cyto-nuclear epistasis. *Ecology* 87:2185–2193.

Chapter 7. Modeling Ant Behavior

1. Franks, N. R., A. Dornhaus, C. S. Best, and E. L. Jones. 2006. Decision making by small and large house-hunting colonies: one size fits all. *Animal Behaviour* 72:611–616.

2. Twain, M. 1976. In *The Higher Animals, A Mark Twain Bestiary*, M. Geisman, Ed. New York: Crowell, 3–6.

INDEX

acorn ants (*Temnothorax,*
Leptothorax, Myrmica): changes
in colony size of, 94; moving to
new nest, 56
adaptive caste distribution, 5, 26–30
age polyethism, 31, 33–37, 150n19
Allomerus decemarticulatus, 123
ant-plant mutualism, 67–68, 82–83,
122–25
Argentine ants (*Linepithema humile*):
adjust searching behavior,
69–71; and aggression and food,
50; colony structure of, 115;
interaction of with other species,
108, 116; and trails, 39–40
army ants (*Eciton, Neivamyrmex*),
17, 82
Azteca spp.: and ant-plant
mutualism, 124–25; interaction
of with neighbors, 105–6;
interaction of with other species,
109, 118

brain and ant colony analogies, 46,
48, 58, 95, 145

bulldog ants (*Myrmecia nigriceps*),
104–5

carpenter ants (*Camponotus* spp.):
and age-related sequence of
tasks, 35; antennal response of to
non-nestmates, 103–4; bamboo
engineering of, 17; cuticular
hydrocarbons of, 52, 102; fungal
parasite of, 112; and memory, 61
Cataglyphis albicans, 150n29
Cephalotes spp., 18
chemical communication, 4–7,
39–41, 148n9. *See also* cuticular
hydrocarbons
colony growth, 75–83
complex systems, 5–13, 19
cuticular hydrocarbons: changes
of, 50, 52, 61, 93, 102; colony
identity in, 50–51, 61, 62; and
food influences on Argentine
ants, 50–51; harvester ant
interactions use of, 51–53;
individual variation of, 64; and
interaction with neighbors,